THE GREEN ORGANISATION

The Green Book
Volume 21

A Work of Reference

Your guide to Environmental Best Practice by companies, countries, councils and communities

www.thegreenorganisation.info

AF887764

Setting the standard

The Green Book is the world's only international work of reference on environmental best practice. It is published by The Green Organisation in tribute to the environmental endeavours and generosity of our Green World Ambassadors.

These are the companies, councils and communities who have won Green Apple Awards for their environmental efforts, and who have taken their commitment to the next level by helping others to help the environment.

They are assisting us in publishing their award-winning papers in The Green Book, and we distribute this valuable educational and informative guide free of charge to environment professionals all over the world.

By helping others to follow their environmental lead and learn from their experiences and successes, they fully deserve the title of Green World Ambassadors.

Not only do they allow us to publish their case histories, but they welcome any follow-up inquiries. Each Green World Ambassador has a contact name and number shown in the Index, and any of them will do all they can to help others help the environment.

The Green Organisation
The Mill House,
Mill Lane,
Earls Barton,
Northamptonshire NN6 0NR
Tel and fax: 01604 810507
www.thegreenorganisation.info

EDITOR'S NOTE:

In the interests of accuracy, these Green Apple Award-winning papers are published here in their original form – basically as provided by the entrants.

As many of the papers are of a technical or specialist nature, it was decided that the experience and expert knowledge of the contenders should not be compromised by the grammatical priorities and consistency of style that usually dominate the editing process.

Our priority is to communicate the aims, methods and achievements of our winners as they wish them to be presented. To this end, the contents have been only lightly edited and we ask for your tolerance of any grammatical shortcomings that might result.

A

ABP FOOD GROUP

WATER SAVING INITIATIVE

Page 44

Despite the business growing significantly over the last few years, the ABP Group has still managed to continue to reduce its CO_2 output. This has been achieved through many strategies but the switch from oil to gas for heating has been a most worthwhile investment.

05274 41444

john.durkan@abpfoodgroup.com

ADVENT PROJECT MANAGEMENT

SEA SALT COTE

Page 47

Back in 1997, David and Alison started selling sea salt they produced on an Aga stove in their country kitchen. Now selling globally in 22 countries, a new state of the art production facility was needed. A £1.25 million project was born and this will provide long term community and economic benefits. The building was designed to be as green as possible, as minimising the energy and carbon footprints of both the building and the production of the sea salt was a must.

01248 719262

ian.williams@adventpm.co.uk

ALBA, BAHRAIN

ALUMINIUM SMELTER TO GREEN OASIS

Page 51

Hydrogen fluoride gas is the most significant gaseous emission from any aluminium smelter, and even small quantities can harm or kill plants. Alba's solution was to develop a lake and plant a forest in the most threatened area, to reduce the risk and the damage. Healthy trees, grass and plants are living proof that the solution is working.

00973 17 837306

salah.aqeel@alba.com.bh

A

ALBERT DOCK, LIVERPOOL (SAVILLS)

GREENER DOCK

Page 56

This large site of historically important buildings suffered from poor waste management and no joined-up strategy to improve matters. When Savills were given the job they wanted to take it from eyesore to eye-catching and they have really delivered the goods. The entire community is committed to the strategy and the performance and appearance are a revelation to all.

0781 3454021
smorrison@savills.com

ATELIER FLEITER, GERMANY

THE ELECTRIC CAROUSEL

Page 57

When the humble childrens' roundabout becomes a practical lesson in power generation, children can learn as they play. This "Edutainment" device not only brings useful power to off-grid locations but it also creates an insight into our electric lifestyles that would be hard to replicate in a classroom environment. In use the carousel powers lights and USB ports on an adjacent lighthouse. This may prompt the children to ask "What is electricity?" and "How does it work?", and although many parents may struggle to answer, it does get everyone thinking about how we take power for granted.

0049 163 3534837
fleiter@fleiter.net

AUTOELECTRO

REFURBISHMENT OF CAR COMPONENTS

Page 59

This company refurbishes and markets car components, particularly alternators and starter motors. This in itself is good recycling as the components are expensive to produce and contain rare earth metals. But Autoelectro go beyond this by ensuring they reuse and recycle throughout the business model. They continue to lead the way in the industry and have earned ISO 9001 and 14001 acceditation.

0127 4656101
tony@autoelectro.co.uk

B

BACK: GABRIELLE
CREATIVE RECYCLING & CLIMATE CHANGE SURVEYANCE
Page 139

BAE SYSTEMS – NAVAL SHIPS
DECENTRALISATION OF COMPRESSED AIR
Page 62

The BAE shipyard at Scotstoun in Scotland operated a centralised compressor system to supply all the compressed air demands of the site. While this set-up has been the industry norm for decades, advances in localised variable on-demand compressors made the old style unreliable and costly. BAE undertook a feasibility study and became convinced that modernisation was the way forward. Since adoption they have been proved right and have made operational savings in excess of £100,000, CO_2 reductions of near 520 tons and a significant reduction in downtime and related costs.

0162 2354774
carolyn.lang@baesystems.com

BEACH SCHOOLS SOUTHWEST
TEACH ON THE BEACH
Page 64

This outdoor-learning company works with schools to take children out of the classroom and give them rewarding lessons in ecology and safety. Even in places like Cornwall, many children do not get access to the shoreline so these unusual lessons leave a lasting impression on young minds.

07824 388079
tess@beachschoolssouthwest.co.uk

BIOGENIE – ENGLOBE
REMEDIATING CONTAMINATED SOIL
Page 67

This cutting edge technology is able to remediate contaminated soil into a fully useable substrate for landworks. As it avoids landfill tax, it offers a saving both to the supplier and to the environment and has diverted 600,000 tons from landfill across the UK.

0118 916 7340
dsmith@biogenie.co.uk

B

BOUYGUES, FRANCE

ENERGY INTERACTION

Page 70

For comparatively low budget and low-use organisations such as schools to look to invest in energy saving investments, it can be very difficult to assess impact and payback. Bouygues has developed a spreadsheet-based system that does all the hard work for them and quickly demonstrates where the investment would yield the highest return.

0207 803 3527

dave.farebrother@bouygues-es.co.uk

BOWERS ELECTRICAL

POWERING A GREENER BRITAIN

Page 72

As testament to the importance of networking at the Green Apple Awards, Bowers have teamed up with previous winner iVol to create a range of cutting edge transformers that could save up to 3.7 million tons of CO_2 if adopted across Europe. The companies are world leaders in efficient transformers, with their products already saving the equivalent of 520 cars' worth of CO_2 per year.

01773 531531

Laura.Gryzelka@bowerselec.co.uk

BRISTOL WATER

SPAWN TO BE WILD

Page 75

The European Eel is a critically endangered species that breeds in UK waters. Many habitat improvement measures have been installed but this project aims to highlight the plight of the eel and gain support from communities in their habitat. By involving children from socially challenged areas, they provide a two-fold benefit for both the eel and the children.

07917 305626

sophie.edwards@bristolwater.co.uk

B

BUCHEON CITY HALL, SOUTH KOREA

AGRICULTURAL TRANSFORMATION

Page 78

This functional and pleasant green space was until recently the site of an abandoned sewage treatment plant. In a collaboration with the council and the community, it has been transformed into parkland and agricultural areas that are divided up between local residents that have no land of their own. The residents are taught how to grow an assortment of crops, and often have enough extra produce to allow them to make an income from their efforts. The site provides a wildlife habitat, an outdoor learning facility and the opportunity for the community to grow valuable produce.

082-32-625-3152
hwa0814@korea.kr

C

CANON EUROPE

25 YEARS OF PIONEERING RECYCLING

Page 84

Canon are recycling pioneers and have offered their customers a recycling route for toner cartridges since 1990. They continue to streamline and expand the scheme, which now serves 24 countries. To date this has saved more than half a billion tons of CO_2 production and removed the need for almost 250,000 tons of raw materials.

02085 888133

pavlin.matia@canon-europe.com

CARILLION MORGAN SINDALL

A1 LEEMING TO BARTON IMPROVEMENT SCHEME

Page 87

This £380 million scheme to improve road links in the north-east involves working in some sensitive habitats such as waterways and woodlands. From the outset the partners established methods to minimise this impact, including relocating trees to a nature reserve, avoiding major works during otter and badger breeding times, and even rerouting the Brough Beck.

078376 11377

steve.gilder@morgansindall.com

CBRE MANCHESTER ARNDALE

MANCHESTER ARNDALE CENTRE ENERGY MANAGEMENT AND SUSTAINABILITY SCHEME

Page 91

The management of the giant Manchester retail centre had given themselves the challenge of slashing £1 million off their utility bills in five years from 2013. To aim this high they needed full support and so trained key staff in carbon management courses. With the new knowledge came a great many ideas that have already taken nearly £250,000 off bills, with many more savings identified.

0161 817 3690

helen.seed@cbre.com

C

CBRE SPEKE

ESTUARY COMMERCE PARK IN SPEKE

Page 94

The site of this commerce park in Liverpool includes three balancing ponds which had become neglected and were prone to visits by poaching anglers. To improve the habitat, appearance and ecology of the park, the ponds were cleared of their abundant reeds, had the larger fish removed, and were given a new lease of life. The move was much welcomed by the ducks and wildlife as well as the park employees.

0151 7093094

john.sloan@cbrems.com

CLARKE: DAVE & SON

RENEWABLE WORKING SHOWROOM

Page 106

CLARKS VILLAGE

WASTE MINIMISATION

Page 97

In an ongoing project to improve recycling rates, Clarks have already taken their figures from 56 to 87%. They have managed this by increasing the recycling options to 12 waste streams, diverting more than 400 tons from landfill and achieving savings well in excess of £10,000.

01458 840 064

aturnbull-kirk@realm.ltd.uk

C

COONEY ARCHITECTS, REPUBLIC OF IRELAND

LEITRIM INNOVATION AND GROWTH CENTRE

Page 100

This project was conceived to refurbish and extend the innovation and growth sector in Leitrim in Ireland and was envisioned to incorporate many sustainable features. Although it fell victim to "value-engineering", the outcome remains an inspiration to users with many passive design principles incorporated. The building features three-storey glazed ventilation stacks that remove the need for forced cooling, and its intelligent insulation means the heating requirements are around one-third of a standard building of this specification.

00353 14533444

patricia@cooneyarchitects.com

COUNCIL OF MINISTERS, KUWAIT

JLEEB CONSTRUCTED WETLAND

Page 196

CWMBRAN SHOPPING CENTRE

EARTH HOUR

Page 103

Forming an Energy Management Forum Group, this 52-acre site has shown that reducing energy and carbon footprint can be achieved. Through innovative commitment to best practices, the centre staff have reduced energy usage, secured sustainable procurement of materials and services for the centre and the wider community and engaged with the community, through events, to educate on the ecological benefits of what they are doing and what the community can do for themselves. Cwmbran can now boast that they have achieved 100% waste going to landfill by segregating and recycling on site.

01633 872121

martyn.haftowski@cbre.com

D

DAVE CLARKE AND SON

RENEWABLE WORKING SHOWROOM

Page 106

Being a previous winner of a Green Apple Award, Dave Clarke & Sons are moving a step closer to achieving their goals set in 2012, when they opened their working showroom. Since then, they have made significant differences by achieving more accolades, more new customers and clients who recognise the importance of what they have to offer. They are continuing to imply and practice their expertise to councils, charities, dealerships and property owners, who are looking to take control of their environment.

01455 618187

sburge@realm.ltd.uk

DOLPHIN SHOPPING CENTRE, POOLE

CHEMICAL-FREE CENTRE

Page 108

A large public space necessitates a large cleaning operation, and this traditionally includes the use of gallons of varied corrosive chemicals. The Dolphin centre in Dorset is the first to adopt the Orbio system, which is an electrically charged saline cleaning method. After trialling it was found to give results equal to the chemical approach while removing tons of CO_2, toxicity and the £12,000 annual chemical bill.

01202 339954

john.grinnell@dolphinshoppingcentre.co.uk

DUBAI MUNICIPALITY

REPLACEMENT OF STANDARD FLUORESCENT LIGHTING WITH LED BULBS

Page 110

The Municipality commissioned Philips to conduct a study at 15 of their premises, with the aim of improving the lighting infrastructure to efficient and sustainable levels. LED luminaires of different specifications provided the solution - reducing electricity use by the value of around £400,000.

00971 4 2063636

arfadhlani@dm.gov.ae

D

DUBAI MUNICIPALITY

SUSTAINABLE WASTE WATER MANAGEMENT

Page 112

Dubai has witnessed significant economic growth in the last 20 years primarily due to its development as a business centre in the Middle East region. Government policies aimed at development of a diversified economy have transformed Dubai city into a fast growing urban region with a large multinational population (about 200 nationalities). Marketing efforts have resulted in the city supporting a arge transient and tourist population. These factors have resulted in the typical environmental and public health challenges associated with growth, namely rising quantities of waste water generated and the need for an ever increasing level and quality of service.

00971 4 2063636
arfadhlani@dm.gov.ae

E

ECO ENGINEERS & COMPANY, SRI LANKA

NEWLY INVENTED BIOUAFF TECHNOLOGY

Page 115

Eco Engineers have developed a system for the treatment of wastewater that negates the need for electricity and generates no sludge. This has significant environmental benefit over traditional systems. Hundreds of these plants have been installed across the world and each one meets or exceeds global standards for treatment while removing many of the environmental issues of their predecessors. The system has the potential to produce a commercially viable amount of biogas that can be captured for reuse as fuel.

031 2225515

eco.engineer@yahoo.com

ECOPAINT, ANGOLA

OCTOPARK

Page 117

Ecopaint Angola's successful Octobuilding, which has a grass rooftop and uses solar panels for heat and light, is a sustainable eco-building that will be part of a new business and technical development complex, known as OCTOPARK. It will represent a business nerve centre for the region and has potential for national and international dissemination. There will be recreational, hospitality, administration and technical support areas that can be used for environmental education.

00244 928113351

geral@ecopaint-angola.com

EDT AUTOMOTIVE

REDUCING EXHAUST POLLUTION

Page 119

The company has developed a device which dramatically reduces the pollutant output from car exhausts. This device, which has been applied to 13,000 vehicles in the UK, reduces emissions by up to 58% on petrol engines and 66% on diesel cars. Customers also report fuel savings of up to 23%.

07772 046576

david@edtautomotive.com

E

EIBE PLAY

SPORT ENGLAND PRIMARY SPACES

Page 121

Aiming to transform over 600 Schools' play provisions, ebie Play and their usual team of installers and sub-contractors, were chosen as the right team to head this project. Most Primary Schools have very little play area and the Primary Spaces fund has culminated in over quarter of a million pounds worth of revenue to provide bright new playgrounds, resurfacing, netting, wall mounts and floor markings, to utilise every inch of space available. Eventually, £150million worth of sporting equipment and facilities will be available to many, developing a sporting habit for life.

01483 813834
rrnapier@eibe.co.uk

EIBE PLAY

WORKING WITH DINOSAURS

Page 124

A pleasure park in Horsham, West Sussex was in a tired state when the council tendered for improvements. Eibe put forward ideas and the council chose them for the project. Inspiration came from an Iguanodon fossil found locally, which is now a central feature of the Dinosaur Park. Most primary schools have very little play areas and the Primary Spaces fund has culminated in more than a quarter of a million pounds being available to provide bright new playgrounds, resurfacing, netting, wall mounts and floor markings. Eventually, £150million worth of sporting equipment and facilities will be available to more than 600 schools, developing a sporting habit for life.

01483 813834
rrnapier@eibe.co.uk

E

ENTERPRISE RENT-A-CAR

20TH BIRTHDAY ROUNDABOUTS

Page 126

To celebrate 20 years of business in the UK, Enterprise took it upon themselves to form a collaboration to transform roundabouts and fields across Britain. They have nurtured and seeded more than three acres of neglected landscape to promote wildlife and aesthetics nationwide.

01784 221387

emma.l.lynam@ehi.com

ENVIRONMENTAL CROP MANAGEMENT

BIO-PESTICIDES

Page 130

ECM are world leaders in the development of safer crop and animal treatments. This project has proved the effectiveness of mustard and garlic-based bio-pesticides as a part replacement for strong chemical applications, with dramatic reductions in pollution, and protecting beneficial insect populations. Their products provide better protection, minimised risk and cost no more than standard chemical applications.

078 31355148

peterclare@dial.pipex.com

F

FLOORBRITE IN ASSOCIATION WITH CBRE AND TfGM

IMPROVE ENVIRONMENTAL AWARENESS AND PROMOTE ALTERNATIVE TRAVEL OPTIONS IN PARTNERSHIP WITH TfGM

Page 132

The aim was to promote viable and sustainable active alternative green travel options for the occupants in a multi-tenanted office development in Bolton, improving health, wellbeing, access to employment and reducing carbon and accessing funds to refurbish the cycle hub and promote cycling as a green travel alternative. In partnership with TfGM, their travel partners and Floorbrite they conducted an online travel survey and based on those results, developed a building travel plan with objectives.

01204 380522

simon.mcgaw@cbrems.com

FSC BLENCATHRA CENTRE

RENEWABLE ENERGY PROJECT

Page 135

This training facility is located in an inspiring but challenging location, 300 metres up the Blencathra slope. Being off-grid for heating has given it high running costs, but a multi-faceted sustainability project has changed all that. Power is now supplied from a new hydro-electric installation over a mile away, and heating is provided by locally sourced biomass. Savings in costs and pollution, as well as the education opportunities, are mounting.

01768 779601

tim.foster.bl@field-studies-council.org

G

GABRIELLE BACK

CREATIVE RECYCLING & CLIMATE CHANGE SURVEYANCE

Page 139

This educational project encouraged thousands of schoolchildren to study current and past local butterfly populations as indicators of climate change. It also had them creating butterfly artwork using recycled items. The studies did show population decline in some species, which raised awareness in the children. The artworks have been combined into a mural, which is on display in the Holy Trinity Church in Coventry where it is helping to raise thousands of pounds for Myton Hospice.

02476 417217

gabrielle@back.F9.co.uk

GIRAFFE INNOVATION

VIRGIN ATLANTIC ECONOMY MEAL SERVICE

Page 141

The humble in-flight meal became the focus of a project to improve both sustainability and customer satisfaction. A key consideration for items in flight is weight as this directly affects fuel use. However, the team at Giraffe did not take this as the only criterion, and built in financial and environmental savings where they could. The outcome is a win for all, with savings across the board – and all made while achieving a 9% rise in customer satisfaction.

07584 350791

t.palmerfry@giraffeinnovation.com

G

GOVARDHAN ECO VILLAGE, INDIA

GEV SYMBIOTIC DEVELOPMENT MODEL

Page 144

As a model to demonstrate sustainable living, this village is a prime example. Rural farming communities such as these can be ideally placed to find alternative avenues for what would usually be waste. They have combined traditional animal power with 21st century solar power to create a community in harmony with its environment. The system is based on the way nature has found a recycling opportunity for every natural waste stream imaginable, and even though some imagination is required for some commodities, they have achieved very close to 100% symbiosis with their environment.

09199 20888350
anandcaitanya@gmail.com

GREEN MOTION

LOWERING CO_2 EMISSIONS

Page 148

This car and van rental company is a pioneer of green travel options, with the lowest combined CO_2 output of any rental fleet in the world. They have sourced their fleet from the greenest cars available to include many pure electric vehicles. They go further by donating to tree planting organisations in the South American rainforest and give greener driving advice to customers.

jonellebradley@greenmotion.com

GROSVENOR SHOPPING CENTRE, NORTHAMPTON

THE GARDEN

Page 151

The Grosvenor Centre houses a secret courtyard garden within its upper floors, but as the surrounding offices are vacant this had been neglected. When new management discovered the area they realised it would be suitable for a team building renovation exercise and a quiet space for busy staff. Hence a low-budget, high-impact project was begun. The outcome is a haven for staff and wildlife with the bonus of free fruit and veg.

01604 637268
russell.x.hall@eu.jll.com

Work of reference on

Environmental Best Practice Volume 21

H

HANG TUAH JAYA MUNICIPAL COUNCIL, MALAYSIA

CARBON MANAGEMENT

Page 155

By implementing a tracking system to show how and where CO_2 is produced, the council in the Malaysian city has been able to start projects aimed at making reductions. This is ongoing but already they have made huge improvements to their output through awareness campaigns and reduction strategies.

00606 2323773

perancangbandar@mphtj.gov.my

HM REVENUE & CUSTOMS (HMRC)

PAPER REDUCTION PROJECT

Page 157

This five-year project aimed to reduce paper use across the entire tax process. By offering users the option to receive and submit electronic forms, by recycling inhouse paper use and a varied assortment of other measures, they have made savings approaching a million pounds and reduced their waste burden by around 10,000 tons.

03000 586851

diane.wailes@hmrc.gsi.gov.uk

HIGH CHELMER

USING PHOTOVOLTAIC PANELS

Page 160

In 2013 High Chelmer installed a solar panel array above a 60,000 square foot Primark store to provide 17 kilowatts of peak power. The installation has proved to be a great success, generating more than 50,000 kilowatts and an income of almost £5,000. Beyond this, it has removed the need for the equivalent amount of mains power and its associated carbon, and has inspired colleagues at other centres to follow their lead.

01245 260755

mick@highchelmer.com

H

HISTORIC SCOTLAND

THERMAL UPGRADE FOR A TRADITIONAL COTTAGE

Page 163

In partnership with Historic Scotland, The Gannochy Trust aimed to deliver a suite of energy efficiency measures on an inter-war period, traditionally built, solid walled cottage. Previous refurbishments had proved costly and resulted in technical issues with condensation, as well as producing excessive quantities of waste and the loss of durable building components. Measured thermal improvements to a solid wall property were achieved with materials technically compatible for a traditional building. The site phase went well with only modest adjustments to the schedule of works and no significant delays. Good indoor air quality was ensured through the use of hygroscopic materials, improved conventional ventilation and the use of existing flues.

0131 6688621

roger.curtis@scotland.gsi.gov.uk

HYDRACHILL

WATER REFILLING STATION

Page 166

Conceived under the *Water for Health* initiative, the HydraChill Water Refilling Station was born of frustration at the lack of access to hygienic mains-fed drinking water for on-the-go consumers. Purchasing bottled water was typically the only option. This inspired them to provide a free and innovative alternative to costly bottled water in high footfall areas. Not only do disposable plastic bottles raise serious environmental concerns, but UK mains-fed water is of a high quality and is fed by a carbon efficient system utilised imaginatively by this company.

01792 863331

nd@hydrachill.com

J

J MURPHY & SONS

WHITE WALTHAM ECO-GARDEN

Page 170

Murphy's White Waltham depot has become an unlikely insect oasis after the company renovated land around the site to form an eco-garden. Using waste materials from the site, they have constructed a reserve that is a welcome home to bees, birds and insects - and a favoured lunch spot for the staff.

0771 2936583
shanelongworth@murphygroup.co.uk

JONES LANG LASALLE (JLL)

MAKING A COMMITMENT

Page 172

The management and staff at Westside Plaza shopping centre have made a commitment to ensure they operate to the best of their abilities with regard to environmental issues and their community. They have had considerable success on these fronts and present a facility that is cleaner, greener and promotes return visits from their clientele.

0131 4423123
westsideplaza@btconnect.com

K

KANNAN: MR C

CHENNAI METRO RAIL PROJECT

Page 210

KELTBRAY

BBC TELEVISION CENTRE

Page 173

To manage waste, you first need to know what you are dealing with. Keltbray have invested in the SMARTWaste software system that helps to give a better prediction of the type and volume of waste a project will produce – even before the first on-site action. This has enabled them to set ambitious yet achievable waste management objectives from project conception.

07711 593116
claire.fundrey@keltbray.com

KELTBRAY

VICTORIA CIRCLE

Page 176

This project took place around some of the busiest streets in Central London, and on first appraisal seemed to require almost 6,000 lorry loads of rubble and backfill in and out of the site. It was decided that by crushing concrete arisings on site, this could be used for backfilling and as base mat, thereby reducing all lorry movements. This has massively reduced local congestion and pollution as well as having a beneficial impact on project overheads.

07711 593116
claire.fundrey@keltbray.com

K

KIER

HI-VISIBILITY RECYCLING SCHEME IN MOLDOVA

Page 179

Recycling opportunities can pop up in the least expected locations. When Kier learnt that a roads and infrastructure project in Moldova was putting residents at risk from road accidents, they used it as an avenue to recycle more than 5,000 items of high visibility clothing. Benefits to both parties are obvious, but the scheme has been expanded to include items such as child car seats and is treated as a springboard for expansion.

0774 0394266

jamie.edmonds@kier.co.uk

KIER CONSTRUCTION

THE CARTERHATCH LANE CONTRACT

Page 182

Kier used their affordable homes project at Carterhatch Lane as a proving ground to test some of their novel environmental initiatives. As the project is a Level 4 sustainable construction scheme it was ideal for trialling three waste reduction measures - the Packaging Return Initiative, Community Wood Recycling and Hazibag. These systems help Kier to achieve their waste reduction targets of -5% per year.

0208 508 5622

mark.fox@kier.co.uk

KIER HARLOW

THE COMMUNITY WAVE

Page 185

Under ever tightening budget restrictions, councils have had to get smarter to maintain their towns and cities. Harlow had a worsening street litter issue, but no budget to apply to its rectification. By making changes to the service delivery system, they have maintained the budget while reducing the indicator value from near 22% to below 3%, beating their own ambitious targets.

07710 582260

mickjenkins@kier.co.uk

K

KM CHARITY TEAM

KENT GREEN TRAVEL CAMPAIGN

Page 187

This charity is an old friend of these awards, and continues to reach heights unimagined 15 years ago when they started a walking-bus scheme at a local school. This project is now used weekly by 40,000 schoolchildren in the Kent area. Having developed a cloud-based data acquisition system, they are rolling the scheme out nationally, which means that the 208,000 car journeys they have already negated will soon grow exponentially.

0844 264 0291

sdolby@thekmgroup.co.uk

KOREA AIRPORTS CORPORATION

LED-ILCMS DEVELOPMENT OF CARBON REDUCTION

Page 189

With increasing pressure from both the environment and the public, air travel is always seeking ways to reduce its environmental burden. Runway lights are a considerable consumer globally and KAC have developed an intelligent LED Individual Light Control Monitoring System that can reduce their electrical demand by 50% or more and significantly increase reliability. Most airports either have all the runway lights on or off, but by installing this system operators can illuminate only the runways and taxi routes that are in use. They can even aim lighting only in the direction required, which saves power and virtually negates pilot errors during poor visibility that have led to collisions in the past.

dalone@airport.co.krl

K

KOREA ENVIRONMENTAL INDUSTRY AND TECHNOLOGY INSTITUTE

GREEN CREDIT CARD

Page 192

The Green Credit Card aims to contribute to reducing CO_2 emissions by offering tangible economic incentives for low-carbon lifestyles and consumption patterns. Furthermore, it is intended to nurture markets for greener products, thus driving eco-innovation to underpin a green economy.

+ 82-2-380-0642
hjlee@keite.re.kr

KUWAIT: COUNCIL OF MINISTERS

JLEEB CONSTRUCTED WETLAND

Page 196

In a desert landscape such as Kuwait, water management is always a primary consideration in construction projects. When a need was seen for a waste water treatment facility, it was conceived as a constructed wetland that would then go on to provide irrigation for a nearby park. This symbiotic development has achieved great success, with huge sustainable and financial advantages over the traditional models of the two schemes.

0965 97404848
ismail.kuwait@aol.com

L

LAND SECURITIES

GUNWHARF QUAYS WASTE MANAGEMENT

Page 200

Large shopping centres produce large amounts of sometimes tricky waste to deal with. Land Securities wanted a more positive approach and they have continually fine-tuned their processes to yield 75% reuse and recycling, and zero to landfill. Having saved more than £60,000, they look to improve even more.

02392 836740
sean.sweeney@landsecurities.com

M

MARINE RESEARCH FOUNDATION, MALAYSIA

PROMOTING AN ECO SYSTEM-BASED APPROACH TO FISHERIES

Page 203

Malaysia lacks an environmental fishery management system and has suffered from significant sea turtle decline as a by-catch of shrimp fishing. In 2007 this NGO promoted a system that prevents the turtles from getting caught in nets. The traditional nets now feature a frontal mesh that turtles cannot pass. The system has gone from strength to strength and is supported nationwide. They have been so popular that they will be mandatory in two years.

00601 98164830

npilcher@mrf-asia.org

MARMAX PRODUCTS

SPECIAL EDUCATION NEEDS RANGE

Page 206

These products for outdoor play at special educational needs schools had key criteria to fulfil before being deemed fit for purpose. By being made from 100% recycled HDPE (mostly from milk bottles) it ticks the eco-schools and sustainable schools boxes, but by being cleverly designed to be safe yet entertaining, it ticks the all-important "happy kids" box too.

01207 283442

jackier@marmaxproducts.co.uk

MR C KANNAN

CHENNAI METRO RAIL PROJECT

Page 210

To not just ensure, but exceed global standards of health, safety and environmental management, the team for the Metro Rail project adopted many of the leading international standards. In setting out this rigid framework for the project they have protected both the environment and the contractors at every stage.

091-766 744 0614

cn_kannan@yahoo.co.in

Index

N

MURPHY: J & SONS

WHITE WALTHAM ECO-GARDEN

Page 170

NISSAN MOTORS GB & NISSAN QASHQAI

NISSAN QASHQAI

See ad on Page 214

One judge said of the Qashqai: "It remains the best small SUV choice, almost regardless of what engine is in place. The 1.5 dCi unit provides excellent economy, with real-world figures closer than you might expect to the official claims."

01923 899933

david.jackson@nissan.co.uk

NOMENCA

BARROW WASTEWATER TREATMENT WORKS

Page 215

The upgrade works to this sewage facility included the replacement of a de-watering system. Nomenca took the option of having the new equipment built and tested off-site in factory conditions - which gave great improvements in local disturbance levels, site health and safety, and equipment quality and reliability. As well as significant savings in time, CO_2 and budget - the project has given the facility a marked reduction in maintenance costs.

01925 281222

samantha.doyle@nmgroup.uk.com

NORTHERN IRELAND WATER

STONEYFORD INTEGRATED CONSTRUCTED WETLAND

Page 219

The village of Stoneyford had been served by a failing sewage treatment facility that urgently needed replacement. Rather than add to the existing facility, it has been closed and NIW land has been used to create a wetland that will take care of the community's waste with minimum investment for at least the next 50 years. This system also creates an amenity for animals, plants and visitors.

07709 309043

Derek.Crabbe@NIWater.com

N

NOTVITCO LTD T/A ORDINARY SKINCARE COMPANY

THE ORDINARY SKIN CARE COMPANY

Page 222

This company operates under self-imposed and very strict rules guiding its sustainability ethos. They have developed a proven effective range of skincare products that have met with worldwide acclaim, yet have done so by sourcing sustainable materials and recyclable packaging. They even design the containers to ensure that every last drop can be used - a lesson to packaging companies everywhere.

01904 347511
ROBERT@ORDINARYSKINCARE.COM

O

O'BRIEN CONTRACTORS

PROJECT DOVE

Page 224

The best-laid plans often fall victim to outside influence such as the weather - as was the case with this project for Nestle. After planning approval. the scheme experienced the highest-ever rainfall, which made the plans all but impossible. O'Brien management brainstormed the issue and came up with a novel work-around that utilised remediation instead of excavation. This not only got the project back on track, but offered a massive reduction in vehicle movements and saved 15 tons of CO_2 emissions.

01926 319724

neilnickolds@obriencontractors.co.uk

@ONE ALLIANCE

CAMBRIDGE GROWTH SCHEME, WATER RECYCLING

Page 228

This project was required to improve and expand a waste water treatment plant in Cambridge. Ambitious targets were set in relation to noise, dust and nuisance - as well as minimising environmental impact. Many of the targets were exceeded, including landfill. This was achieved by reusing excavations on site and using recycled plastic items where available.

07880 405115

carol.hardingham@skanska.co.uk

O

OSTLERS CIDER MILL

CIDER VINEGAR

Page 231

Aiming to be certified organic – with four acres of orchard and a further four dedicated to wildlife – managing this is no easy feat. With a solar system and wood burner in place, they are self sufficient; and with the help of a mixture of sheep and horses, the land is fertilised organically - which in turn helps the wild flora and fauna from turning to turf. The orchards produce a unique product "Vinegar with the Mother" that can only be purchased in independent shops and has been used on BBC 2's Great British Menus' Finals Week.

01271 321241

info@ostlerscidermill.co.uk

OXFORD BROOKES UNIVERSITY

LOW CARBON OXFORD

Page 232

With more than 20,000 staff and students, the transport carbon footprint of this university is potentially huge. By investing in euro-6 buses - using KERS systems designed by the Williams Formula One team - they have managed to reduce this by about 1.4 million car journeys per year, a saving of almost 1000 tons of CO_2 annually.

01865 488238

HRUSSELL-EMMERSON@BROOKES.AC.UK

P

45 PALL MALL VENTURES LTD

45 PALL MALL

Page 234

Having only recently taken on the management of this building, The team found that there was no structured waste management plan in place. By quickly making the right decisions they have already got to grips with the issue and are delivering up to 75% recycling rates on a daily basis.

0207 2698618
alan.donnelly@eu.jll.com

PETERBOROUGH HIGHWAYS (SKANSKA)

ONE PLANET LIVING

Page 236

Delivering a sustainable highways maintenance contract, covering all aspects of One Planet Living, was the aim here. Health and wellbeing; carbon and waste reduction; sustainable transport; and all manner of environmentally friendly improvements were top of the agenda. In the first year, 97% of waste was diverted from landfill; 40% of energy was produced with the introduction of solar PV panels; and electric cars were chosen over petrol or diesel ones.

0787 6874444
alex.harwood@skanska.co.uk

PROJECT JANSZOON, NEW ZEALAND

CONSERVATION AND WILDLIFE PROJECT

Page 240

Project Janszoon was established to implement a new concept of collaboration with the Department of Conservation, whereby philanthropists commit to fund an ecological transformation of a part of the conservation estate against agreed and defined indicators, and the Government commits to secure those ecological gains, once achieved, for the long term.

006435287804
robyn@janszoon.org

Work of reference on

Environmental Best Practice Volume 21

R

RAININ INSTRUMENT, USA

TERRARACK

Page 245

Laboratory waste is rarely useable for recycling due to the chemical and biological risks of its contents. With this in mind, Rainin have developed TerraRack which - despite being robust enough for any pipetting operations - is also collapsible, recyclable and yet uses 50% less plastic than the standard racks on the market. The system has been designed to be easy to use, compatible with any size pipette tip on the market and is 100% recyclable PET so does not need breaking down when it goes for recycling.

510-564-1697

geracar.nervis@rainin.com

RENAULT UK/DACIA DUSTER

DACIA DUSTER

Page 247

One judge said of the Duster: "I've chosen it because it offers unbeatable value." Another said: "As well as being great value for money the Duster is actually a pleasant car to drive and economical too."

01923 697592

JEREMY.TOWNSEND@RENAULT.CO.UK

RG GROUP

DORRIDGE DEVELOPMENT

Page 248

As the main contractor for this retail development, RG Group aimed to regenerate the town centre. Added bonuses of this development were utilising rain water harvesting, trialing of waterless urinals and hybrid generators and deliver a carbon neutral project with minimal waste. In total 78 tonnes of waste was prevented by reuse within the local community (saving 4.5% of demolition waste and preventing waste to landfill rising by 335%). With this also came a new biodiversity area and a school tree planting programme was born, with 1800 trees planted.

07951 125390

colorado.goldwyn@rg-group.co.uk

R

ROCHDALE EXCHANGE SHOPPING CENTRE

CAR PARK EFFICIENCY

Page 251

One of the often overlooked avenues for savings in public buildings is the car park. Rochdale Exchange targeted this as their latest sustainability project and have achieved considerable CO2 and financial savings as a result. By switching to LED lighting, closing the virtually unused upper levels and installing a ticketless system, they have also made it much more user-friendly.

01706 764961
chris.doidge@rochdaleexchange.co.uk

ROYAL PRIORS SHOPPING CENTRE, LEAMINGTON SPA

WASTE MANAGEMENT INITIATIVE

Page 253

Addition of waste streams inevitably leads to subtraction of general waste weights. Royal Priors found this to their benefit when they introduced food waste recycling into their portfolio. They have increased recycling of food by over 32 tons and have improved the output of other streams too.

01926 450150
les@royalpriors.com

S

SANLAMERE UK

SEA CHANGE BATHROOM WATER USAGE

Page 255

So often the simplest ideas are the best. Sanlamere considered water use in the typical bathroom and realised that sink waste water is ideal for flushing toilets. By designing an all-in-one unit they easily capture sink waste which is then held until required for flushing. This can save up to 70% of water use and offers ongoing savings to the end-user.

0208 5448091
maria@sanlamere.co.uk

SCOTTISH ENVIRONMENT PROTECTION AGENCY

NETREGS

Page 259

The Netregs website was originally set up in 2004 to help SME's understand their environmental obligations and to simplify compliance. Since SEPA took it on they have enhanced it still further with such features as a self-assessment tool, video clips and e-learning programmes that give SME's a one-stop-shop to improve their performance and make savings as well as avoid incidents and prosecution.

Their current development plan looks to improve this still further with updates and new content to give the end user an enhanced experience and easier searching. Improvements so far have given savings of more than £1-2000 per business, 5000 new subscribers and a saving to SEPA in costs related to phone calls they receive regarding related issues.

01896 754797
Gillian.Bruce@SEPA.org.uk

S

SKANSKA

NORTH WEST CAMBRIDGE DEVELOPMENT

Page 263

This 150 hectare new build community project sees the development of some 3000 new homes, academia, sports facilities and all the related infrastructure. Skanska are laying out the site and have included some cutting edge sustainability measures to include the largest rainwater recycling plant in the UK and a huge underground bin system - removing the need for some 9,000 wheelie bins. Despite all the works, local wildlife will also benefit from the development with improved habitat.

07764 655791

sean.carter@skanska.co.uk

SONNE ENERGETICOS, MEXICO

CARBON FREE COLLECTIONS

Page 267

This company collects waste cooking oil for conversion to biofuel or for use in chemical applications. They take their position as the largest in Mexico seriously and act sustainably wherever possible. Trees are planted so their customers operate carbon-positively and by using the biofuel in their own trucks they effectively close the loop.

081 838 88300

afhuglerv@gruposonne.com

SOUTHWAY HOUSING TRUST

GREENING OUR GREY CITY

Page 268

After a request from disabled there was an opportunity to enhance the area with respect to appearance, green-space and flood mitigation - while giving their tenants the access they required. By creating a grass reinforced mesh driveway and replanting the surrounding land, they have no tenants with difficult access to their home, Southway achieved far more than the original brief demanded.

0161 448 4200

p.reece@southwayhousing.co.uk

S

SUBANG JAYA MUNICIPAL COUNCIL, MALAYSIA

GREEN COLLABORATION

Page 273

This Malaysian council aims to achieve Green City status for Subang Jaya by the year 2030. Their strategy for this covers many fronts including the Green Collaboration initiative. This neighbourhood collaboration scheme partners families with available land for them to tend and grow fruit and vegetables for their own use and for resale. The council also supports landowners to do the same with their plots and promotes people with neglected gardens to do likewise. Families have achieved complete self reliance from their efforts, which in turn reduces shopping trips and thereby congestion and CO2 production. Nearly 400 families with no garden of their own now manage plots on public land, and this figure is increasing by 30-40 every month.

00603 80263125
yan_upuu@yahoo.com

Index

T

TEAM VAN OORD

MEDMERRY MANAGED REALIGNMENT

Page 277

Coastal realignment and coastal sea defences are becoming a sustainable option now, due to rising sea levels and other economic challenges. Team Van Oord have unique and innovative approaches to these issues and have influenced the Environment Agency on future policies. With this project, they have protected 348 properties, sewage works and caravan parks which equates to £90 million of economic benefits, and given Green Tourism opportunities to grow.

01273 492212

Jacqui.bandy@mackley.co.uk

THE CRYSTAL

THE CRYSTAL

Page 280

To demonstrate the important role that technology plays in creating a sustainable future, The Crystal was purpose built in 2012 and is the first building in the world to achieve the highest ratings in both of the world's accreditation bodies, LEED and BREEAM (Platinum and Outstanding respectively) and holds numerous other accreditations for its excellence. To date, it has welcomed over 200,000 visitors and over 75,000 students and constantly receives customer feedback of over 90% to the good. They have also published three books on sustainable living, in conjunction with global experts.

0207 0556407

Samantha.white@siemens.com

TOSHIBA TEC UK

E-STUDIO 306LP MFP

Page 284

The e-STUDIO 306LP MFP combines with the e-STUDIO RD30 for removing toner colour from printed documents. As a result, paper usage can be reduced to one fifth while providing support copy and print speed of 30 prints per minute.

014915 70970

ashley@henley.co.uk

T

TREETOPS EXECUTIVE RESIDENCES, SINGAPORE

DISCOVER GREEN PROJECT

Page 286

Although this is a leisure destination, that does not reduce the management's desire to promote green living. They have installed many energy-saving features and they promote these ideas to their visitors. Their own programme of waste and energy reduction continues to yield results as well, such as a 50% saving on electricity in seven years. They have achieved this by using the most efficient fittings, by applying cutting edge integrated heat recovery systems and with an expansive water recycling operation.These and other measures yield a CO_2 saving of more than 1,500 tons annually.

00 65 68870088

zelene@treetops.com.sg

TRUSTFORD

GOOD CORPORATE CITIZEN

Page 291

As a major brand car retailer, this company has many showrooms and high related costs. They set out a strategy to reduce costs, improve sustainability and invest in technology that would help them to achieve these goals. By installing BMS systems, teleconferencing equipment and low energy lighting among a raft of other measures, they have already reduced their CO2 footprint by 18%.

0207 025 6598

Lara.Edwards@redconsultancy.com

U

UAC BERHAD, MALAYSIA

ENVIRONMENTAL INITIATIVES

Page 295

The company is committed to reducing its carbon dioxide emissions and they have so far achieved up to 35% reduction in some aspects of this ongoing project. This has also produced cost savings and reduced electricity consumption.

00605 2912033
TAN_KEAN_LEONG@UAC.COM.MY

ULSTER SUPPORTED EMPLOYMENT

TO ACHIEVE BS 8555 LEVEL 3

Page 298

This all-inclusive improvement project was employee led, as the staff felt more could be done to enhance their sustainability efforts. By adopting the BS8555 standard, it formalised their approach and ensured all gains were quantifiable. This has been proven effective with energy efficiency up by 27% and a 60% improvement in their recycling performance.

02890 356600
tfuller@usel.co.uk

V

VAUXHALL MOTORS

BIODIVERSITY & WILDLIFE IMPROVEMENT & PROMOTION PROGRAMME

Page 301

Vauxhall's Ellesmere Port site is much more than just a massive car plant. The management has formed a collaboration with the Wildlife Habitat Council and Wirral Wildlife Trust and together they have created and managed woodland and meadow areas of significant natural importance. As well as creating safe havens for many important plants and insects, they have seen Peregrine Falcons raise brood after brood on site.

07930 852129

elliot.bramall@vauxhall.co.uk

VECTAIR SYSTEMS

ENVIRONMENTALLY FRIENDLY AIR FRESHENER

Page 304

Air fresheners in public facilities are products that are more likely noticed when they are not installed, but when used they can have VOC, recycling and waste implications. Vectair are major global suppliers and wanted to improve the sustainability of these products, so they invented a recyclable and biodegradable passively dispersed system that uses no propellants or electricity - yet performs as well as any other system.

01256 319500

cwonnacott@vectairsystems.com

VICTORIA HOUSE PROPERTY PARTNERSHIP

VICTORIA HOUSE

Page 307

Management at this site idntified opportunities to improve their wastes management and recycling rates, and so have invested in improved recycling provision and training for staff. Having previously improved their efforts to 70% recycled waste they set - and have already met - the current target of 80% with the promise of more to come.

0207 269 8618

alan.donnelly@eu.jll.com

V

VOLKERFITZPATRICK

WRABNESS EMBANKMENT STABILISATION

Page 309

This railway embankment improvement project has been on the drawing board for ten years, but due to difficulties of the working environment it had never got started. As the risk of ground-failure increased, it became necessary to complete the works - and some novel planning was required. By working with the custodians - the RSPB - Volkerfitzpatrick completed the build with no detrimental effect to the woodland and estuary areas, and they have ensured that the area should be virtually maintenance-free for the next 100 years.

0782 4599353
james.mcmorrow@volkerfitzpatrick.co.uk

VOLKERHIGHWAYS

LB HACKNEY CIVIL ENGINEERING TERM CONTRACT

Page 312

The team at VolkerHighways set out to be exemplars of best practice during highway construction. Of paramount importance was accessibility for cyclists, pedestrians and wheelchair users during project works. They used varied methods to examine and enhance their performance, such as employing "mystery shopper" techniques and cycling inspections.

075 00064826
Jason.Convey@volkerhighways.co.uk

W

WYKE FARMS

WYKE FARMS' 100% GREEN PLAN

Page 316

Aiming to be one of the greenest brands in grocery history, Wyke's 100% Green plan is backed up by promises to significantly reduce their carbon footprint by producing all of their energy from solar and biogas sources. All of this can be found on their website (the only dairy company to do this) and can be monitored too as time goes on. They are now 100% self-sufficient and have saved over £2 million by reducing waste and packaging costs, and saved over 40 tonnes of CO_2.

07887 780 464
jodie@connectedpr.com

APB FOOD GROUP

WATER SAVING INITIATIVE

PROJECT AIM

The aim of the project was to significantly reduce water usage at ABP Cahir. The project highlights ABP's commitment to water stewardship with a fully integrated approach.

As a food processing facility, the site uses a large amount of hot water so the site management focused on reducing hot water use and thereby reduce natural gas consumption.

An extensive water reduction programme was launched in January 2014. Good housekeeping provided initial water reductions. LEAN manufacturing principles were integrated into the new projects to provide solutions to counteract significant water users.

Extensive sub-metering was installed to measure and monitor significant water users. A cost analysis was carried out to highlight the "true cost" of water. This helped the site to finance the project and focus attention on important cost saving initiatives.

The project has prepared the site for the European Water Stewardship (EWS) assessment. It also helped ABP Food Group to achieve recertification with the Carbon Trust Water Standard.

PROJECT ACHIEVEMENTS

A significant reduction in water demand was achieved. The site reduced water usage by 50% in 2014 compared to 2013. This reduction in the site's water usage reduces the hydrogeological impact on the aquifer from which ABP Cahir abstracts. This has been analysed and tested as part of the EWS assessment.

Similarly, a reduction in water usage decreases the risk of contamination to the accepting surface water body following waste water treatment.

Reduced water usage directly correlates to reduced electricity consumption to pump, treat and dispose of water. Reductions in hot water usage (45°, 65°, 90° and steam), reduces the site's natural gas consumption and as a result reduces the site's carbon footprint.

A reduction in the amount of chlorine and salt used to soften and treat the abstracted water was also achieved. Approximately 10% of treated waste water is recycled for use in the lairage, truck wash, yards and WWTP which also reduces the site's water abstraction.

WHAT DID THE PROJECT INVOLVE DOING AND WHY DID YOU DO IT?

- The volume and cost of water at each temperature (cold, 45°, 65°, 90° and steam) was established. Significant water users were identified and extensive sub-metering was installed to analyse their water usage. KPIs were developed for each significant water user to help monitor and control water usage
- Sensor operated hand washers replaced manually controlled taps. (45°)
- Timer operated knife sterilisers and sensor operated saw sterilisers replaced continuous flow sterilisers. (90°)
- Improved wash-down practices across the three food production facilities on side were introduced. Optimised pressure and correct nozzle sizes were also vital in reducing water consumption during wash-down. (65°)
- Tripe washing was fully automated with adequate timing to meet strict processing criteria. (90°)
- The tray washing process was fully optimised to significantly reduce water usage. (90°)
- The water flow to the evisceration table was reduced from continuous flow to now match the movement of the processing line. The water now flows for 15 seconds out of every 70 second time period. (65° and cold)
- Heat recovery from the refrigeration system was installed to preheat process water. This further reduced the site's gas consumption by approximately 15,500 kWh per week.

ABP Food Group has established a sustainability agenda where the mission statement is "Doing more with less". The project was undertaken to primarily reduce water usage and fossil fuel consumption. As a result, the site has improved its business performance through reduced costs, while increasing production.

Accurate information on the "true cost" of water was established. Therefore, a capital investment programme in water reduction initiatives and/or technologies can now be based on accurate information.

Staff culture changes have been implemented with a greater respect for water saving throughout all of the facilities.

WHAT DID IT COST?

A case for water and natural gas reductions was put forward to the board of directors. A decision was made to finance the project in stages. Positive results provided further financing. The accurate cost analysis further provided funding for the project.

Good housekeeping was key to developing these projects which is obviously a very cost effective way to reducing water usage. All of the sterilisers in the abattoir were replaced at a cost of approximately 25,000. Sensors and timers were inexpensive. The sub-metering and associated costs were less than 10,000.

A relative reduction (m3/tonne) in water usage of 45% was achieved. Therefore 45% less water was abstracted and 45% less water was treated in the WWTP.

Similarly, a relative reduction (kWh/tonne) in electricity consumption in the WWTP of 13% was achieved. Also, increased retention time within the WWTP improved the plant's performance and hence decreased the risk of emissions to the environment. A relative reduction (kWh/tonne) in natural gas consumption of 35% was also achieved.

Cost saving of over 100,000 was achieved over the year.

WHO AND WHAT HAS BENEFITED?

John McLaughlin, General Manager at ABP Cahir, has enthusiastically spearheaded the reduction initiative.

In particular, abattoir management and staff have implemented the reductions to process water usage. Most of the 90° water reductions were achieved due to the abattoir staff.
Engineering/maintenance staff are fully aware of the importance of the reductions and have provided solutions for many of the technical issues.

Effective communication with staff via screens and information boards to show progress and help understand initiatives have also been critical in driving the reductions and instilling a sense of ownership.

WAS THERE ANYTHING INITIATIVE ABOUT THE PROJECT?

ABP Cahir will be the first ABP Food Group site to undergo assessment for the EWS Standard. We aim to be the first company in Ireland and the first food processing site in Europe to achieve this goal.

ABP Ireland is a member of the EPA-funded Community of Practice in Water Management, which is a multi-industry steering group. ABP Cahir has hosted one of the recent round table meetings. This group provides a platform for discussions and offers solutions to water management issues.

Water is not a finite resource and these initiatives have been proven to significantly reduce water usage. This has led to an enthusiasm to further reduce resource consumption, not just at ABP Cahir, but across the entire group.

WHAT ARE THE LONGER TERM BENEFITS?

The extensive sub-metering will lead to continuous improvement with tighter targets and benchmarking. Industry standard suggests continuous flow in all sterilisers and the evisceration table is critical. The site's initiatives have all passed strict food safety standards and have generated significant savings counter to these industry standards.

These practices are currently being rolled out to all Irish sites. The systematic approach to water reductions and problem solving are being shared with colleagues through the LEAN programme. The directors are keen to replicate the significant reductions at all ABP Ireland sites.

Excellent communication and data presentation with all stakeholders was achieved by presenting water reductions graphically throughout the site. Feedback was also supplied through ABP's website and the company's internal social media. A Sankey diagram was also developed to help staff understand water consumption throughout the site.

A top level buy-in was a fundamental part of driving the initiatives. One of the key elements to reducing water anywhere on the site was to break down the problem. Therefore LEAN was an important tool. Measurement and verification were also key.

Further reductions are planned in the carcass wash (45° water), cold tripe washing, cold wash down to replace hot wash down (in specific areas) and condensers in the refrigeration plant (cold).

Further sub-metering is required for greater in-depth information. This will allow the site to set more KPI's and targets to fully optimise water usage. These KPIs can be used to benchmark all of the Irish sites.

The water and fossil fuel reductions at ABP Cahir have been independently verified by SGS, on behalf of Origin Green, as part of ABP Ireland's sustainability charter. ABP Food Group was recertified to the Carbon Trust Water Standard for the second time, which included a site visit to ABP Cahir where another independent investigation was carried out. ABP Cahir was awarded the OSI Environmental and Sustainability Awards 2014 for water reductions.

ADVENT PROJECT MANAGEMENT

NEW SALT COTE

PROJECT AIM

Since starting to sell sea salt in 1997 when it was first produced on an Age stove in their country kitchen, David and Alison Lea Wilson dreamed of building a state-of-the-art facility to produce their "culinary gold."

The home-grown business moved into temporary portacabins and took on extra employees but David and Alison quickly realised that the rapid growth of their business meant that they needed more modern permanent facilities to support the high demand for this luxurious sea salt enjoyed now in 22 countries worldwide.

The business is situated adjacent to the shores of the beautiful Menai Straits in a location designated an Area of Natural Beauty. This required a sensitlve approach to the design of the new building in terms of complying with planning regulations.

It has taken ten years to fulfil David and Alison's dream and this was mainly down to securing the right building shape, size and layout, fighting economic conditions and raising sufficient funding to allow the project to happen.

The building also meets the strict demands of food hygiene regulations as well as the hlgh standards of the retailers that sell the salt.

PROJECT ACHIEVEMENTS

The £1.25 million project wlll see Halen Mon provide long term community and economic benefits to the island and the region well into the future.

The new salt cote, which has become a landmark destination in south Anglesey and north west Wales, supports the local community and the economy as well as preserving the 16 permanent jobs and supporting the local suppliers, customers and delivery companies.

Worth its salt: the new building

The project was realised by using a mix of support from a private owner, the bank, the lottery's Coastal Communities Fund and a variety of public sector grant funding.The new building provldes a long term and sustainable business base as the previous production was carried out in portacabins on a temporary planning consent basis.

New facilities were an urgent necessity. By incorporating quick forms of construction such as a steel frame with timber integrated wall panelling finished off with a Welsh larch cladding and steel roof, the provision of a low maintenance and energy saving building was a key factor.

By receiving free issue Welsh larch, the project assisted Bangor University's Bio-Composites Centre with an innovation research project to review effects of kiln drying the larch.

WHAT ARE THE ENVIRONMENTAL BENEFITS?

From the initial concept the building has been designed to be as green as possible by using energy efficient design principles embedded and ensuring a minimum energy/carbon footprint from both the building and the sea salt production.

For example, the external cladding is kiln-dried Welsh larch and the CO_2 footprint is reduced by use of heat recovery mechanical ventilatlon to ancillary areas, spatial future-proofing for heat recovery off process extract, a boiler fed from process LPG supply, variable volume process extract ventilation and energy-efficient lighting and lighting controls.
The building has achieved Energy Performance certificate Rating B and uses a Building Management System to control equipment. A log-burning stove is used in the reception/visitor shop and it is fuelled from a locally sourced and managed woodland.

WHAT WAS INNOVATIVE ABOUT THE PROJECT?

Worthy of note is the zero direct energy input utilised during the brine production and heating process. The building is simple and elegant and uses steel stud and timber structurally integrated wait panels with insulation system for cost, speed. air tightness and sustainability.

M&E expert consultants familiar with condensation treatment from swimming pools and other food production projects were utilised and the client also invested in Carbon Controls expertise to audit current and proposed energy performance and efficiency and develop proposals and recommendations.

The made by hand production process can best be described as energy intensive. The production equipment and process associated with manufacture of the 'raw product' namely brine, both the equipment and process (involving low temperature boiling under vacuum. inverier frequency control of compressors/pumps, friction derived brine heating) is judged by Carbon Control to represent an exempiar interns of energy efficiency. In the old building resultant moisture-laden warm air was extracted via roof mounted fans and exhausted directly to atmosphere.

A waste heat recovery system has been introduced together with a Building Management System to assist to monitor energy usage and inform in order to improve the reduction of energy and waste as time goes by. A visitor display screen allows visitors to see the energy saving results being achieved.

The team has worked on this project for more than ten years and the husband and wife owners of Anglesea Sea Salt have been diligent in expanding their business both in the UK and overseas and never gave up on their dream to build a new state-of-the-art production facility. There has been strong leadership from senior management matched by input and commitment from the entire workforce eg assessing existing energy performance of production process thereby assisting the carbon and M&E consultants to develop robust energy monitoring for the new production facility to allow energy management, control and reduction to be embedded as 'business as usual'.

WHO AND WHAT HAVE BENEFITED?

The new salt Cote and Visitor Centre is a destination magnet for south Anglesey. It has already been voted Best Visitor Attraction 2015.

The island relies heavily on tourism and the fact that Halen Mon is known as a famous welsh produce and a global-reaching brand means that visitors will now be welcomed in a warmer and nicer environment. This will add to he economic benefit of the visitors and local people can have a guided tour of the production facility to see the salt being harvested and also taste the different varieties of salt being produced.

Seeing salt being made by hand is a unique and educational experience and is already proving to be of great interest to local schools and other educational providers. During the course of the project life approximately 30 local contractors and/or consultants have been used. The new Salt Cote will assist to make the facility a pleasant place to work for the staff.

WHAT ARE THE ECONOMIC BENEFITS?

In addition to saving costs during the bulld project (eg via a Bangor University Research project) we obtained free issue Welsh larch cladding which completely encapsulates the building other than the roof. Energy cost reductions are achieved through investment in new production facilities and robust energy monitoring which enables the business to remain competitive and improve profitability and thereby re-invest profits in growlng the business and with its exports across the globe. Stock-taklng costs have already been reduced due to the deslgn of the new facility .

Before this project. salt production was undertaken in portacabins. The salty and humid environment meant that these portacablns were not adequately insulated, were leaking energy and humidity direct to the atmosphere and quickly coming to the end of their life. This brand new facility has achleved an Energy Performance Certificate Rating of B (Appx C) which is a substantial improvement. The score of 45 is at the top end of the B scale and almost achieved rating A.

The building was only finished at the end of January 2015 and therefore no long-term data exists to demonstrate sufficient evidence of impact. However it is clear that already substantial energy improvements and reduction in waste are being realised with the advent of the Building Management System, waste energy recirculation and new methods of worklng following the Introduction of new working practices following the visit to a factory using LEAN methodologies.

ARE YOU PLANNING ANY FURTHER DEVELOPMENTS?

The design has catered for the possibility to retrofit a wood-fuelled biomass boiler to provide heating and hot water generatlon in the future. This would reduce LPG consumption and the environmental impact as fuel would be locally sourced and sustainable.

The clients expect to deliver further significant cost savings in labour and energy as the building was designed following visits to America and studying a Japanese factory. The staff had a two-day tour of the Toyota factory where they witnessed LEAN principles in action.

CAN OTHER FIRMS REPLICATE THESE FEATURES?

The facilities are capable of being fully replicated by others and the design of both the building and the mechanical and electrical concepts can be used for the creation of other buildings. The energy monitoring and control approach can also be involvemented by other businesses.

The team and client intend to make the most of the building as a visitor centre and an exernplar Building Information Modelling (BIM) project.

ALUMINIUM BAHRAIN (ALBA)

ALUMINIUM SMELTER TO GREEN OASIS

PROJECT AIM

- Achieve zero effluent discharge by stopping discharge of 1278 million litres per year of wastewater effluents to the sea
- Stop abstracting groundwater and thus conserve 2574 million litres per year of precious groundwater, an important non-renewable resource of the country
- Achieve annual reduction of 5% in wastes disposal.

PROJECT ACHIEVEMENTS

The HRH Princess Sabeeka Oasis is a source of pride for ALBA. It tells future generations of ALBA's success in transforming a desert into a beautiful oasis.

Desert blooms: ALBA has transformed the Bahrain landscape

One of the major goals of the forest and lake was effluent recycling, necessary to achieve zero effluent discharge. The lake was conceived as a means to gather and store the entire surplus sweet water effluents that were discharged to the sea until then.

Hydrogen fluoride gas is the most significant gaseous emission from any aluminium smelter. The presence of even very small concentration of hydrogen fluoride gas, will make plants sick, and some species may die.

The wind direction in Bahrain is predominantly to the south so the south side of ALBA will receive the highest level of emissions.

The decision to set up a forest at the south side of ALBA was a brave and challenging decision of senior management.

The healthy grass, plants and trees that are flourishing in the forest clearly demonstrate the environment friendly operation of ALBA. This living data is more convincing than any amount of measured statistical data on Alba emissions.

The oasis is divided into four sections:

Water cover

The water cover is divided into three sections. The Artificial Lake is one of the largest lakes in the Kingdom of Bahrain, with an area of about 10,000 square metres and a capacity of more than 22 thousand cubic metres of treated water. The water resulting from the cooling process of the aluminium moulds is pumped into the lake and sent to the oasis for irrigation and for other agricultural purposes.

The lake has a diverse aquatic life such as fish, ducks, turtles, frogs and freshwater fish such as the Tilapia and Veral fish. The lake is a haven for migratory birds such as flamingos, hoopoes and gulls.

The waterway, which covers an area of about 1,500 square metres, is considered an important part of the oasis. It carries the downstream water from the Casthouse and was designed with the purpose of cooling the industrial waste water to the appropriate degree, as well as purifying it before using it for the lake and other irrigation purposes. The third part of the water cover is the Natural Lake, which covers about 3,500 square metres. Shaped like the old traditional natural water wells in Bahrain, the water is mainly used to irrigate plants.

Green cover

The HRH Prince Sabeeka Oasis has a long green patch of more than 300 metres and contains many species of plants and more than 15,000 trees, including fruit trees such as apple, mango, pomegranate and banana together with Sesbania, Cordia, Oleander and Palm trees.

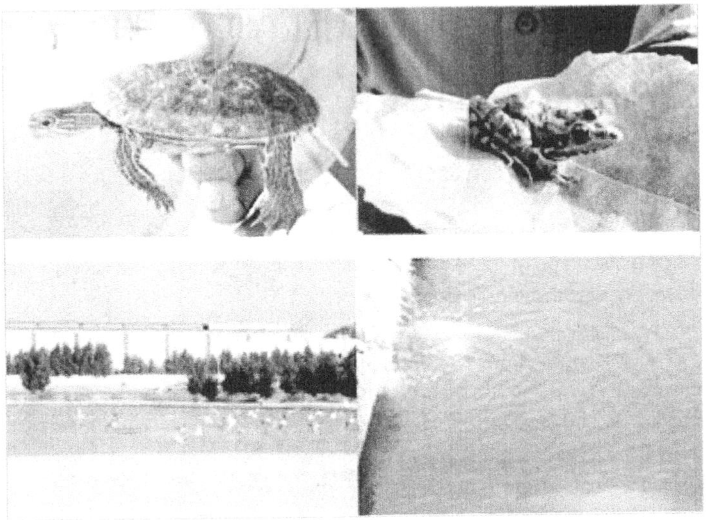

Jump to it: making their home in the lake

Oasis farm

The Oasis farm contains many types of wild animals that ALBA co-operated with Al Areen Resort to make this natural farm become reality in June 2014. Animals released in farm included deers, ostriches, chickens and goats and few other types of animals. Food for the animals is from the canteens' waste and from vegetable waste.

Vegetable garden

The vegetable garden covers about two hectares and contains many types of vegetables and Bahraini local fruit trees. Several irrigation methods are used such as the drip irrigation, along with the traditional method of sweetwater immersion used traditionally in Bahraini farms.

Various types of vegetables are planted in the garden for every season and the Landscaping Section in ALBA distributes the seasonal crops, which are estimated at more than six tonnes per year, to ALBA employees.

WHAT DID THE PROJECT INVOLVE DOING AND WHY DID YOU DO IT?

Alba's approach to the founding and development of the HRH Princess Sabeeka Oasis was not simply a matter of going green but it was to express and demonstrate the interest of an industrial plant in environment and agriculture.

Through the oasis, ALBA communicates to the society its commitment towards the environment and showcases its achievements in agriculture. ALBA regularly conducts environmental and agricultural tours for guests from various organisations, as well as many visits from schools' and other establishments. The company also participates in environmental events such as the Bahrain International Garden Show.

Extending the company's efforts towards community services, ALBA has undertaken many initiatives including a forestation plan to transform the desert area of Askar Road into a green patch of more than 8,000 square metres and other similar projects as part of the restoration of the Al-Malikya Beach.

This beautiful green space within ALBA reflects the company's success at various levels, which are not limited to the production of high-quality aluminium alone. All these have been achieved through the support and guidance of top management and the unlimited support from employees.

Reaching a major milestone in the pursuit of environmental goals, ALBA is all poised to achieve zero effluent discharge from its aluminium smelter and power station plants. Through another landmark environmental achievement, the scarce groundwater resources in Aquifer B in Bahrain will be ensured to last many years longer, thanks to ALBA making alternate sustainable arrangements to meet its sweet water requirements.

The ALBA forest with its lake plays a key role in achieving and sustaining these noble environmental goals. These objectives are set from the voluntary commitment of the Management towards sustainability through proactive environmental improvement.

WHO AND WHAT HAS BENEFITED?

In November 2005 ALBA commissioned an eight km water pipeline at a cost of five million US dollars between its calciner plant and the main smelter plant in order to achieve these environmental goals.

The pipeline was laid to discontinue the operation of RO plants at the aluminium smelter site that produce sweet water using the groundwater as feed.

It not simply a matter of going green but it was to express and demonstrate the interest of an industrial plant in environment and agriculture.

Reaching a major milestone in the pursuit of environmental goals, ALBA is all poised to achieve zero effluent discharge from its aluminium smelter and power station plants.

Through another landmark environmental achievement, the scarce groundwater resources in Aquifer B in Bahrain will be ensured to last many years longer, thanks to ALBA making alternate sustainable arrangements to meet its sweet water requirements.

The RO plants at the aluminum smelter use natural groundwater as raw material for producing sweet water. Natural groundwater in Bahrain is far less salty, compared to seawater. Therefore it is a valuable natural resource.

This is a non-renewable and limited natural resource and so it is very important to conserve it for future use. Depletion of non-renewable and limited natural resources is a major environmental issue globally.

The Water Desalination Plants at the Calciner plant are eight km away from the aluminium plant, use seawater as raw material for producing sweet water. Desalination is achieved by utilizing the waste heat recovered from the calcination of petroleum coke. The volume of water produced far exceeds the ALBA plan requirement and hence a major part of this potable water is injected to the national water network

By bringing water produced from seawater at the Calciner plant and by establishing a 20-acre forest and an effluent recycling lake in 2004 for irrigating this new forest, ALBA aluminium smelter plant has achieve zero effluent discharge, which is a major environmental achievement for any industry.

The RO Plants generate large amount of salty water as effluent, which is discharged to the sea. Being salty it cannot be used for irrigation. All the other effluents from ALBA are sweet water effluents, which can be used for irrigation after purification.

By constructing the lake and forest, all such sweet water effluents are recycled and do not go to the sea any more. Now by stopping water production at RO plants, generation of the saltwater effluents is stopped, making it possible to achieve zero effluent discharge from the smelter and power plants.

The forest was designed and planned to achieve several environmental goals. In the winter of 2005, when the migratory birds from far away Siberia settled in their hundreds at the lake in ALBA forest, the dream of providing a natural bird haven at the barren ALBA desert was fulfilled.

The environmental perspective about this forest is that the forest and lake along with the reed bed develops naturally in to a self-contained natural unique eco-system in Bahrain.

This forest-lake eco-system will have birds, small animals, butterflies, migratory birds, chicken, ducks rabbits, fish, turtles, frogs and aquarium fishes, while trees, plants and grass will provide green cover.

There should not be any external input to this ecosystem except water, agricultural chemicals and electricity for irrigation pumps.

The rainwater run-off from the first rain in any chemical industry could be contaminated with chemical dust that settled on the roofs and yards during the year.

One solution to address this issue was to build a large pond to collect the first rain. The lake at ALBA serves this purpose also. The rainwater from a large portion of ALBA is directed to this lake through a two-metre wide drain more than one kilometre long built specifically for this purpose. The water collected is used for irrigation and thus it is a water conservation programme as well.

Freshwater lakes easily grow algae and fill it up, making it a very unpleasant sight, in addition to creating serious operational problems. This was a major concern during the design stage of the lake. The crystal clear water in the lake that is free from any algal growth is proof of the success of design.

The lake was planned to have fish that grows on its own without feeding any food. This requires good quality water free from pollutants and sufficient nutrients. The reed bed of water plants was designed to purify the water and provide nutrients for the fish.

The water plants can absorb many harmful chemicals if they are present, and at the same time this natural growth automatically provides nutrients required for the fish. So in the lake a lot of fish can always be seen at the place where the water falls into the lake from the reed bed, in order to eat incoming nutrient matter from the reed bed.

Flora and fauna are equally important in forest ecosystem. Many species of fauna are naturally attracted to this forest and make it their home. Birds, butterflies and other smaller organisms have arrived this way.

Aquarium fishes normally survive in high purity water maintained in controlled conditions. Aquarium fish no more than five centimetres long were introduced in the lake as an experiment. These fishes have grown to 50 centimetres or more. Some 40 of them are estimated to be present in the lake now.

In this way another objective of the forest to demonstrate the high purity of Alba effluents was achieved. The forest will provide the necessary area for using this water for irrigation.

Of particular difficulty was the effluent from vehicle workshop coming out of the oil separator. Occasionally it could get contaminated with traces of oil. Therefore a secondary oil trap was built at the entry of the lake for these effluents.

The vegetable garden in the forest further demonstrates the good air quality. Vegetables grown here were tested along with same species of vegetables from the market at Bahrain University and other labs for those chemicals that are likely to contaminate the vegetables at ALBA farm. The results showed that the vegetables from ALBA and from the market were similar and safe.

The air quality over the forest is continuously monitored using open path monitoring technology. The concentration of hydrogen fluoride, NO_2 and SO_2 gases are continuously monitored and recorded.

The beauty and aesthetics is an important factor in the success of the design. Normal plantation would have been planting trees along straight lines on flat land. In order to provide a pleasing and natural look, a hill was introduced. Tress plants, shrubs and green cover were planted in a mixed manner without forming any geometrical pattern, so as to achieve a natural appearance.

The decision to set up a forest at the south side of Alba was a very brave and challenging decision by senior management. The healthy grass, plants and trees that are flourishing in the forest clearly demonstrate the environment friendly operation of ALBA. This living data is more convincing than any amount of measured statistical data on ALBA emissions.

albasmelter.com
Launch into Future

ALBERT DOCK, LIVERPOOL (SAVILLS)

GREENER DOCK THROUGH ENVIRONMENTAL IMPROVEMENT

PROJECT AIM

Our long term aim is to eventually introduce a complete waste compound for the whole estate where full control of segregation, bailing of cardboard, food waste etc. can be even more carefully managed.

Our goal as an estate is to achieve as close as possible a "Zero to Landfill" approach. With the buy-in now of all occupiers at Albert Dock we want to be a fully environmentally friendly tourist attraction. This recent progress is simply first stage as we see it.

PROJECT ACHIEVEMENTS

To bring together some 60 separate businesses and 2,500 staff members, including several public sector bodies, and still gain an understanding and a collective drive to improve the waste management of such an iconic location has been our biggest achievement so far.

The improvement to the look and feel of the estate is amazing. To only have one collection company as opposed to eight is also something we are proud off at this early stage of our progress. We hope to improve continuously.

WHAT DID THE PROJECT INVOLVE DOING AND WHY DID YOU DO IT?

When Savills were instructed as the managing agent at the Albert Dock in Liverpool two years ago, I was very quickly given the support I needed to instigate changes to the waste collection and recycling procedures, which I had identified as major issues across the estate.

The dock's 19th-century warehouses, which are the largest single collection of Grade I listed buildings in the UK are now home to shops, bars, restaurants and museums and galleries, and residential apartments.

For many years, tenants and occupiers had been able to make their own arrangements for waste collection ,which resulted in a lack of control, overflowing bin bags left along roadsides and very low levels of recycling.

We decided that a uniform approach would be the most effective course of action and began a period of consultation with the board of directors, tenants, landlords and residents over the course of many months.

We then approached B&M Waste Services and procured a single-fee-per-bin deal, which actually offered savings to the majority of occupiers.

WHO AND WHAT BENEFITED?

Within 12 months, 95% of tenants had changed their practices and B&M were appointed across the whole estate. This was a remarkable achievement given that each leaseholder, tenant, occupier and resident did and still pay for their waste collection services individually.

The level of commitment and drive needed from all 2,500 residents and employee's at Albert Dock was incredible. Individual staff training and re-training began in each venue, with some starting again from scratch. Working closely with B&M Waste Services resulted in attitudes beginning to change dramatically.

During the same period we also identified the absence of a convenient recycling facility on the estate as being an issue and subsequently created the first waste and recycling compound since Albert Dock's refurbishment in the 1980s.

This has allowed us to collect cardboard from many of the units to increase levels of recycling even further. Three years ago around 40% of tenants were recycling at best, but now that is up to around 93% and improving all the time.

The efforts and dedication of so many individual business to work together for a common environmental goal has been brilliant, and one we are all proud of here at Albert Dock, Liverpool.

ATELIER FLEITER

THE ELECTRIC CAROUSEL

PROJECT AIM

On the one hand, THE ELECTRIC CAROUSEL brings power to powerless places; places that have restricted connection to an electric infrastructure. On the other hand, it is an invitation to start thinking, both in a playful and a very practical way, about the physical phenomenon of electricity, its generation, storage and consumption, which is nothing less than the core energy of modern society.

What exactly is electricity? Where does it come from? How can it be stored? What does 100 watts mean? What do they feel like? These are questions only a few adults can answer, let alone kids.

This project aims for old and young to start looking into the subjects of renewable energies, energy harvesting and decentralized power generation, which gets more and more important to deal with if we want our kids to grow up within healthy living circumstances.

But THE ELECTRIC CAROUSEL does not aim to educate with a raised moral finger, but indeed presents an unusual solution for an everyday problem – charging mobile phones – in an exploring and interactive way. Edutainment at its best.

PROJECT ACHIEVEMENTS

The first project installation was put into action in September 2014 as part of the Think blue campaign of Volkswagen AG on a children s playground at the resort of Westerland on Sylt. THE ELECTRIC CAROUSEL can not only complement an existing electric infrastructure, but also provide electricity to locations where otherwise there would not be any.

It achieves a decentralised energy generation independent from fuels, unpredictable natural forces as well as predictable human forces such as maintenance contracts among others.

The carousel can be operated easily by four-year-old children and puts out electricity at very low rounds per minute. As security always comes first, all of the components are protected by several security measures against wanted and unwanted mechanical overload. The electric system is secured against unauthorised access and can be reached with specialized tools only (for regular maintenance). The cables connecting the carousel with the lighthouse are run underground.

The electric system is based on 12 volts low voltage and can generate up to a maximum power output of 10 amps. Not much, but enough to charge up to four mobile phones simultaneously with running the lighting effects of the lighthouse and the internal operating as well as the energy-monitoring system.

WHAT DID THE PROJECT INVOLVE DOING AND WHY DID YOU DO IT?

THE ELECTRIC CAROUSEL seems to be a usual merry-go-round at first sight, but when set into motion by kids playing with it, a miniature lighthouse right next to it lightens up. But the highlight feature of the lighthouse can only be seen when looking more closely: it provides four USB-power outlets to be used for charging mobile devices, usually smart phones.

The power for illumination and charging comes straight from a generator within the carousel by underground cable. This is a win-win situation for kids and their parents and a concrete value added on top. If more power is generated than used or when it is time for the kids to go home, a rechargeable battery stores surplus energy.

In case the carousel is not being used for a long time, which is quite unlikely to happen, two photovoltaic cells on the lighthouse ensure the batteries do not be completely lose their charge.

WHAT DID IT COST AND WHERE DID THE MONEY COME FROM?

The carousel was realised with the investment of Volkswagen AG (VW). As it was part of the Think Blue campaign in connection with the Kitesurf World Cup 2014 held on Sylt, VW was the exclusive supporter of this project as well as the tenth anniversary of the kitesurf championship. Therefore, the overall colour design of the carousel and the lighthouse matches the campaign. The carousel was officially inaugurated by none other than Robby Naish, an important player in the international wind- and kite-surfing scene. The total costs were 45.000 .

Now, how did this project come to life? VW had already been in contact with Atelier Fleiters' THE ELECTRIC HOTEL. An employee did know Sebastian Fleiter from one of his talks about self-made power generation.

VW approached Sebastian with the idea of building a device that "is fun and linked to electricity". He came up with a merry-go-round but intended to build it himself with the help of his team of engineers. But as the approval for public spaces implied many specific standards and licences to be permitted by the German TÜV, Atelier Fleiter started its search for a partner who could deliver already certified components and be open to innovation.

This was found in Kaiser&Kühne Freizeitgeräte, an internationally renowned provider of playground equipment. Kaiser&Kühne Freizeitgeräte built a merry-go-round and modified the substructure to integrate the generator unit built by Atelier Fleiter.

Regarding the lighthouse, a whole electronic system including power infrastructure as well as a computer interface were developed to provide information on energy-monitoring and upload data into the internet.

After a development and construction period of six months, THE ELECTRIC CAROUSEL was finally assembled on the children's playground of the resort of Westerland, Sylt.

WHAT IS INNOVATIVE ABOUT THE PROJECT?

The innovation lies within combining freely available modules and components and fusing those different parts of technology and design in order to develop an electronic system that makes it possible to transform movements and muscular power into electricity. Atelier Fleiter has specialised in communication concepts following the claim of self- made electricity. Various other projects with similar systems of electricity generation by muscular power have been realised. They range from tiny soap-bubble machines for small children to STROMODROM, a large-scale 45 metre-long slot-car racing track powered by generator-bikes up to a mobile water power plant. Last but not least is THE ELECTRIC HOTEL, a mobile communication power plant.

The idea of an electrified carousel comes from the same mindset: to bring the idea of a decentralised and sustainable power generation closer to the public. Decentralized energy generation is indeed very beneficial for communities in every nook of this planet.

There are still third world countries that hardly have running water or electricity, but mobile phones are to be found in every single corner of the globe. THE ELECTRIC CAROUSEL shows one way of generating self-made electricity via muscular strength and movement, which is there in everyday life anyway.

There are many other possibilities for implementing such a system – small or big scale. It is a system that is independent from fuels, unpredictable natural forces as well as predictable human forces such as politics and money business to name only a few.

WHAT ARE THE LONG-TERM BENEFITS?

A long-term benefit besides raising awareness would be the adaptation of such a system to other places and corners of this world. Enhancements and further developments can achieve a system of higher scale and thus, more people; people that are truly in need, can benefit from it.

THE ELECTRIC CAROUSEL shows that energy generation, even only to a small amount, can be easily achieved with better preserving nature, living conditions and our surroundings. It only takes us to find a way to capture, store and connect the energy generated by even our everyday movements in order to be used when we need it.

Atelier Fleiter is currently working on a way to connect its various devices to smaller power grids of island systems so that an emergency power structure can be established in order to maintain at least communication possibilities.

AUTOELECTRO

REMANUFACTURE OF AUTOMOTIVE PRODUCTS

PROJECT AIM

Autoelectro is involved in the remanufacture and supply of automotive components specialising in starter motors and alternators. We purchase scrap and end-of life parts and also sell our products on an exchange basis so we get the old failed part back whenever we sell a new one.

These parts are then used as the raw material and brought back into an 'as new' condition through a manufacturing process that involves stripping each part down to its bare components, replacing any wear assemblies such as bearings and brushes etc, and thoroughly testing other components and only re-using if they meet strict criteria.

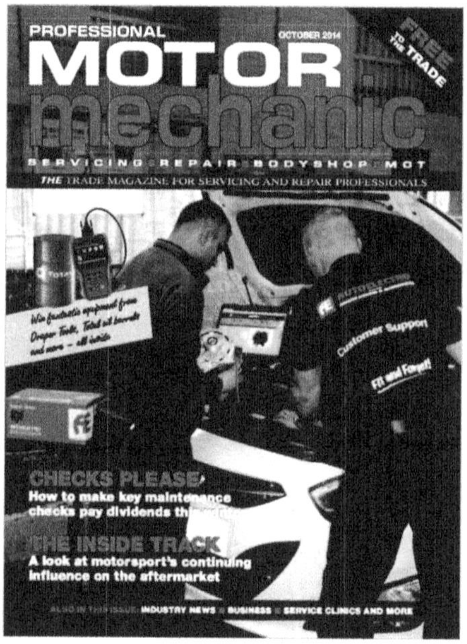

In the picture: Autoelectro parts

The components are then re-assembled and the finished product tested before being packed and warehoused ready for re-sale. In many instances this remanufactured part is superior to the original part since any flaws in the original design will have been rectified.

PROJECT ACHIEVEMENTS

Autoelectro is now probably the largest independent remanufacturer and supplier of starter motors and alternators for cars and light commercial vehicles in the UK. Our products are supplied throughout the country and Southern Ireland.

We have invested heavily in the latest test equipment allowing us to remanufacture parts for the newest vehicles where the technology includes computer controlled alternators and starter motors that fit the 'stop-start' applications. This has resulted in us having the best range and availability. Our products and processes have had a significant positive environmental impact since previously manufactured components have been reused, saving on raw materials and built-in energy in the original manufacturing process.

We have a company ethos of being environmentally friendly, old boxes are used for packaging, lights are on movement sensors etc. and we have recently been awarded ISO14001 certification in addition to ISO9001.

WHAT DID THE PROJECT INVOLVE DOING?

Our business is based on recycling and bring back into service faulty and old starter motors and alternators for cars and light commercial vehicles. There is a significant demand for these products but the complexities of the newer designs have made them more difficult to remanufacture and requires significant skill and investment.

We are perhaps the largest of only a handful of companies involved in the remanufacture of these parts and, as a result, we have had to source components and machinery from all over the world.

IS THERE ANYTHING ABOUT THE PROJECT?

We will shortly be taking delivery of a new computerised test bench for testing alternators. This bench captures test data from working alternators so that we may then use this

for final testing our remanufactured since test data is very difficult to obtain. The continuous investment gives us a marketing edge over our competition.
Over the years we have spent several hundreds of thousands of pounds in our facilities and we now occupy some 35,000sq ft of floorspace. Much of this investment has been achieved by continual reinvestment of profits.

Our services allow customers to purchase original equipment quality products at a competitive price allowing people to keep their cars on the road in addition to having significant environmental and sustainability benefits. The company is now ISO9001. ISO14001 and OHSAS18001 accredited and we supply most distributors of aftermarket car parts in the UK. Autoelectro is an ethical business and has never had an enforcement action taken against it.

BAE SYSTEMS – NAVAL SHIPS

DECENTRALISATION OF COMPRESSED AIR

PROJECT AIM

The aims of the project were to:

- Review the centralised compressed air system feeding the Scotstoun yard pipeline and evaluate the efficiency
- Understand the overhead and capital cost implications of the then current centralised compressor setup
- Develop an outcome that will offer a proposed strategy for compressed air going forward considering both the future short term/long term utilisation of the yards.

The project outcomes were a proposal for reduction in utility usage and cost, identifying savings relating to reduced maintenance costs, and to realise efficiency savings by identifying waste.

PROJECT ACHIEVEMENTS

- Reduced capacity – the system was reduced from 450 litres to 100 litres per second max
- Cost saving – £101,000 per annum
- Reduction in electricity usage – annual saving of 765,960kwh, smaller units that are variable rate and area specific meaning that they are running to cover specific times
- Reduction in carbon footprint – annual saving of 518 carbon dioxide tonnes
- Increased reliability – positive impact on production uptime as the number of leaks has reduced
- Reduction in maintenance costs – smaller proportion of yard pipeline pressurised, leaks are easily identifiable.

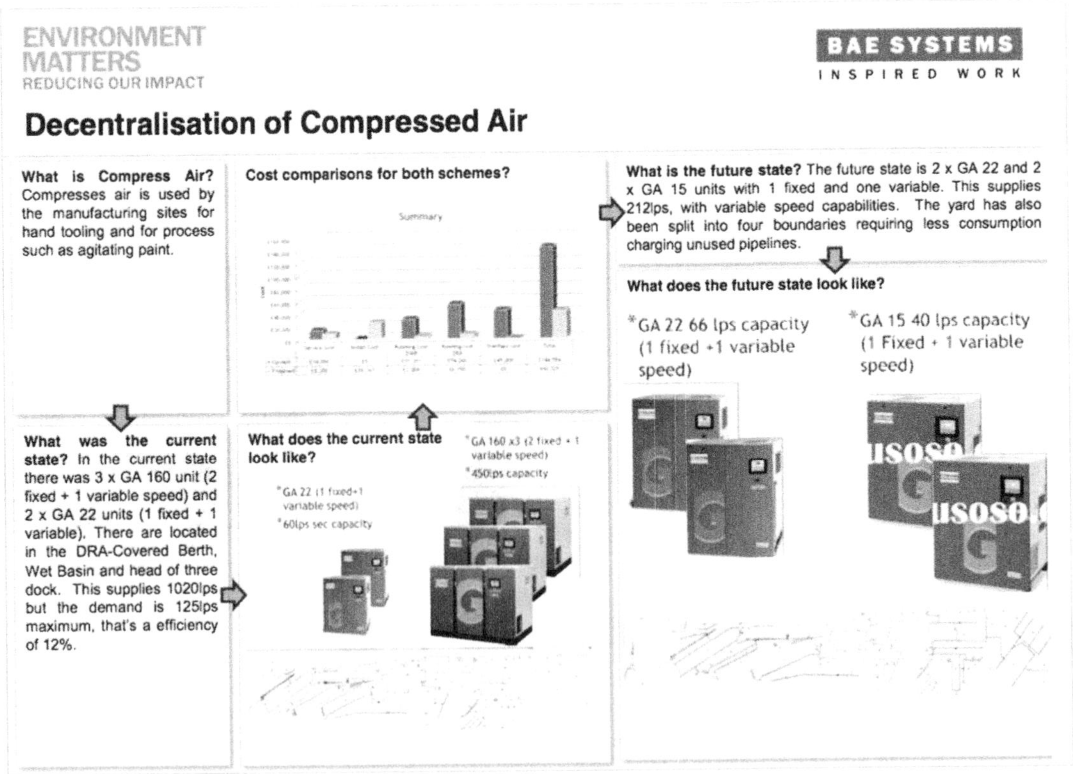

WHAT DID THE PROJECT INVOLVE DOING AND WHY DID YOU DO IT?

Under our commitment to be an environmentally responsible and efficient business as demonstrated by both our ISO 14001 and ISO 50001 management standards, we seek to identify areas of significant usage where we can improve efficiency, compressed air being one of these.

The perceived need for carrying out the project was to understand the overhead and capital cost implications of the current centralised compressor set-up, with an outcome that will offer a proposed strategy for compressed air going forward considering both the future short term/long term utilisation of the yards.

Additionally the current compressed air set-up consists of large compressors working in series over the entire yards.

There are a number of air leaks that are either untreated or inaccessible. The large travelling distance of the systems requires continual running to maintain pressure, exasperating both the run times and running cost.

The project was also carried out to deliver cost avoidance of purchasing new equipment caused by over burdening the kit and reduced pressure due to leaks in the system.

The decentralisation of compressed air project followed the principles of lean manufacturing and the associated tools and techniques. Using these techniques allowed engagement of stakeholders and the demonstration of a different mind-set towards improvements to the system.

The first step in the project was to undertake a leak detection survey of the system and repair any leaks that were discovered. The supply to unused areas of the site was then isolated, an outcome of this was the segregation of the yard into two sections. The two sections of the compressed air line was data logged for a week to establish the actual demand on the compressors.

After the demand was established the new compressed air system could be sized appropriately. At this point in the project, additional opportunities were taken to install variable rate instead of constant rate compressors to closely match demand, the compressors were also put on the building management system (BMS) to allow close control.

WHO AND WHAT BENEFITED?

Economic

- Cost savings – £101,000 forecasted annual savings
- Increased reliability – positive impact on production uptime as the number of maintainable parts has reduced along with fewer air leaks
- Customer value – competitive advantage gained through efficient savings. Site of areas previously affected by service interruption has been reduced as individual compressors control smaller areas.

Environmental

- Reduction in electricity usage – annual saving of 765,960kwh. This is enough to power 160 houses a year
- Reduction in carbon footprint – Annual saving equivalent to 518 carbon dioxide tonnes
- Graphical breakdown of financial savings.

ARE YOU PLANNING ANY FURTHER DEVELOPMENTS?

The project will be rolled out at our Govan site, where stage one is currently under way with a leak detection survey complete and identified leaks being addressed.

When all leaks are rectified, the true demand for compressed air on site and the schematic network of compressed air can be developed.

BEACH SCHOOLS SOUTH WEST

TEACH ON THE BEACH

PROJECT AIM

This project aims to educate as many children as possible about protecting the environment. Beach Schools South West is an award-winning outdoor learning company dedicated to getting children out of the classroom and onto the beach for curriculum-linked learning and to reconnect them with nature

School's out: outlook is bright for learning on the beach

A staggering 81% of children in the West Country did not go to the coast last year according to Natural England's January report (2015), despite nowhere in Devon being more than one hour from the beach and nowhere in Cornwall being more than 17 miles from the coast.

Beach Schools' sessions reconnect children with the environment in which they live, help them learn to learn, help to create the marine custodians of the future and help to teach them safety at sea and on the shoreline.

PROJECT ACHIEVEMENTS

Beach Schools' sessions have proven positive outcomes and help young people who otherwise find it difficult to engage in the classroom to learn curriculum-linked subjects in the natural environment. We have an unashamedly bottom-up approach to educating these children about the natural environment and work happily in partnership with 'top-down' marine educators such as the Marine Biological Association, The National Marine Aquarium and The Blue Mile.

In our first six months of working we put more than 500 children through outdoor learning. We teach on the beach and are determined to live up to our mission statement to give every child access to learning on the shoreline.

Children who access Beach Schools sessions have their vision of the world around them permanently and positively changed. We have given dozens of children a chance to experience the marine environment who have never, ever been there before.

It's estimated that 95% of primary school children in Plymouth have never been to the beach. We have already changed that for the better.

This year we have accessed a small grant which means that this summer we will give 750 seriously disadvantaged children free, fully-funded Beach Schools sessions. We are very proud indeed of that.

Our work ties in with Ofsted Outdoor Learning best practice at a time when the government is actively encouraging schools to get children learning outside. We are very proud to be reconnecting children with natures.

WHAT DID THE PROJECT INVOLVE DOING AND WHY DID YOU DO IT?

The project involves helping schools, community groups and organisations to get children out of the classroom and onto the beach for outdoor learning activities/studies – particularly those who find it difficult to engage in the classroom.

We are a not-for-profit community interest company based in Devon. Outdoor learning helps children engage with learning, to learn to learn. Our method of pedagogical learning is child-led and ties in with Ofsted Outdoor Learning best practice.

We would like to see every child in the South West and beyond have access to learning on the beach. Beach Schools South West can provide outdoor learning sessions for all ages from three to 16s.

We have qualified Beach Schools leaders who are first aid and safeguarding trained. They are also DBS checked every three years.

All of our practitioners are passionate about teaching the children in the outdoor environment and work hard to deliver fun and educational lessons on the shoreline. They are fully mobile and can deliver a lesson anywhere that they have permission to do so.

Outdoor learning has proven positive outcomes and is a really great way for children to learn without realising it. Many children are not taken to the beach by their parents.

There are some astounding statistics here in the south west that say that 95% of children did not visit the beach (Natural Connections 2014)' and a more recent study published in January of this year by Natural England states that 81% of children in the west country did not get to the beach last summer.

Our project aims to encourage these children to visit the beach with their schools and their families. The project will also help them to learn about the marine environment, how to look after it and to educate the marine custodians of the future.

This project was set up as a Community Interest Company, a not-for-profit social enterprise. So far it has been launched on a shoestring by Alexis Bowater and Tess Stuber, who between them have five children aged eight and under.
Beach Schools South West was a new take on Forest Schools with the added element of the marine environment thrown in. Both Alexis and Tess are passionate about getting their children outside and can see the benefits of it and decided that it was the right thing to do.

To encourage others out onto the coast line to explore, to den build, to toast marshmallows on a camp fire, to return children to the wild - we've heard it a lot recently, because it works!

WHAT DID IT COST AND WHERE DID THE MONEY COME FROM?

The cost so far has been minimal, with no real overheads as they are working from home. Alexis and Tess have invested approximately £3,500 between them and rely heavily on funding from charities and organisations. They have received some grants in the last six months.

WHO AND WHAT HAS BENEFITED?

The first season of trading (summer 2014) saw 500 children/young people benefit from outdoor learning on the beach. This year they aim to at least double that figure and thanks to funding from a local Devon charity have been able to give 750 free beach schools places to schools from disadvantaged areas of Devon.

They work on the project together tirelessly promoting and working towards delivering sessions on the beach for children from disadvantaged areas.

They also take bookings for weddings and parties, and for other organisations such as schools and holiday clubs. This project will become as sustainable as Forest schools, which has already saturated the outdoor learning education sector.

ARE YOU PLANNING ANY FURTHER DEVELOPMENTS?

Confidence is building in the innovative Beach Schools ethos and Beach Schools South West will grow beyond the South West of England and be recognised nationally as the outdoor learning company that delivers beach environmental and educational activities, tying in with the new national curriculum and providing environmental studies for all ages.
There are plans to become a training centre for those teachers that wish to develop their skills and be able to teach on the beach independently, which would be a real asset to any school/organisation.

There is already a site that we would like to be able to utilise and are working towards this in partnership with a local surf life-saving club. Beach Schools South West has enormous success waiting in the wings and we are confident that the impact of its benefits will draw more and more people to get involved.

Children will benefit from the lessons and will have improved concentration and engagement in the classroom. They will be healthier as a result of being outside and they will be more considerate of the people around them and the world in which they live, they will have increased vocabulary due to learning in a different environment.

Emotional and physical well being are improved and children and adults alike are generally happier!

Last year saw Beach Schools South West win the regional Green business award at the Venus Women in Business awards and in January this year went on to scoop the National Green Business Award.

BIOGENIE-ENGLOBE

REMEDIATING CONTAMINATED SOIL

PROJECT AIM

The aim was to divert hazardous soils and other suitable material from landfill void. We wanted to offer a sustainable solution for these waste streams and at the same time give the client a cost effective alternative to the traditional disposal methods.

We entered into a joint venture with Biffa waste services whereby the treated material is beneficially reused on their landfill sites for restoration purposes.

PROJECT ACHIEVEMENTS

Last year alone, more than 100,000 tonnes of material that would have been destined for landfill void was treated by us and beneficially reused for restoration across our three UK sites.

As well as the huge environmental impact, we saved our clients a substantial amount of money as our process is landfill tax exempt.

Following establishment of the first UK soil treatment facility (STF) in 2005, Biogenie-EnGlobe, has grown to become the operator of the largest number of UK STFs. Through this network of treatment facilities, a joint venture with Biffa waste services, we are able to provide sustainable, fully compliant and crucially, landfill tax free, off-site treatment of a wide range of hazardous wastes.

WHAT DID THE PROJECT INVOLVE DOING?

We have set up three facilities on Biffa landfill sites where there is a need for restoration materials. These facilities are at Skelton Grange (Leeds), Redhill and Meece (near Stafford). The facilities have a substantial amount of plant involved as the process involves drawing air through hydrocarbon contaminated material to help speed up the natural process of degradation and effectively clean the soils/material in a matter of weeks.

We are able to treat a range of materials from soils to track ballast, gully waste and de-littered road sweepings. Initially, we need a site investigation report, a chemical analysis or a sample for us to analyse. If none of these are available, we visit site and assess the material.

A typical analysis would need to include:

- TPH (total petroleum hydrocarbons)
- PAHs (poly aromatic hydrocarbons)
- Metals, pH and SVOCs/VOCs as required (semi-volatile and volatile organic compounds).

If we receive the above information along with an estimated volume, we can provide a quotation within 24 hours of the enquiry. Once we receive the material, we will conduct a reception analysis, then place the material in a batch with similar material and commence the treatment phase.

We use a variety of treatment methods to remediate the material to a level where it is acceptable for restoration purposes on landfill sites. Typically, the material is ready for beneficial reuse within 8-16 weeks depending on the level of contaminants.

WHY DID YOU DO IT?

The cost of hazardous waste disposal has increased year on year since 2004, impacting on the overall costs for countless projects across the UK. There was a need in the UK to divert as much material as possible from landfill void as well as a driver from producers of C & D waste to source sustainable outlets and also avoid landfill tax.

By partnering with Biffa, we also have a guaranteed outlet for the treated material and there is minimal transport involved as all of our facilities are based on the landfill sites that we are providing with restoration material.

WHAT DID IT COST AND WHERE DID THE MONEY COME FROM?

The facilities are jointly funded by Biffa and Biogenie-EnGlobe and are based on technology developed by our parent company EnGlobe Corp. Costs so far for a total of five facilities so far are in the region of £3m. (Two sites have since closed as the need for material has ceased).

IF QUANTIFIABLE, WHAT DID THE PROJECT ACHIEVE IN TERMS OF SUSTAINABLE DEVELOPMENT, ECONOMY, ENVIRONMENT AND/OR EQUITY?

To the end of 2014, we have treated 1,095,184.39 tonnes of material that would otherwise have ended up in landfill void. All of this hydrocarbon contaminated material was treated and used for restoration on the adjacent landfills which due to the proximity substantially cut down on transport and double handling.

WHO AND WHAT BENEFITED?

Aside from the environmental impact of landfill diversion, our clients have benefited from use of a fully compliant solution for what had been a waste that was only suitable for landfill. The savings for our clients are substantial as the process is completely landfill tax free. Biffa also has a reliable source of clean material for their restoration schemes. Our clients are a mix of developers, contractors and local authorities that have built on brownfield or contaminated land.

WAS THERE ANYTHING INNOVATIVE ABOUT THE PROJECT?

Up until our first facility opening in 2005, no one had attempted to treat hydrocarbon impacted soils in the UK. Our bio-treatment process has been extensively developed and used in Canada by our parent company before moving into the UK and we are continually investing in R & D to find ways to make the process even more efficient.

CAN OTHER ORGANISATIONS BENEFIT FROM IMPLEMENTING YOUR METHODS?

Anywhere there is a brownfield development there will be contaminated land. Any organisation that needs to dispose of this kind of material can benefit from using our facilities and make major cost savings as well as the obvious environmental advantages.

ARE YOU PLANNING ANY FURTHER DEVELOPMENT?

We are planning to open a further three facilities in the next two years and are investing researching additional complimentary treatments such as thermal desorption to treat the more contaminated sludges.

UK WIDE COVERAGE
LANDFILL TAX EXEMPT
FULLY COMPLIANT AND SUSTAINABLE SOLUTION AVOIDING LANDFILL VOID
PARTNERSHIP WITH BIFFA ENSURES TREATED MATERIAL IS BENEFICIALLY REUSED ON THE ADJACENT LANDFILL
HAZARDOUS SOIL TREATMENT FACILITIES
CONTACT US NOW TO SEE HOW MUCH YOU COULD SAVE ON YOUR OFF-SITE HAZARDOUS MATERIAL DISPOSAL COSTS
GREEN APPLE AWARD WINNER 2015
GREEN WORLD AMBASSADOR 2016
HELPING OTHERS TO HELP THE ENVIRONMENT
BIOGENIE AND BIFFA
PROUD TO BE GREEN WORLD AMBASSADORS
Biffa
Biogénie
enGlobe
sales@biogenie.co.uk
0118 916 7340
www.biogenie.co.uk

BOUYGUES

ENERGY INTERACTION

PROJECT AIM

Bouygues Energies & Services provides a range of energy services to clients in the private and public sectors and is a framework contractor on the REFIT programme.

Two of the most popular and commonly undertaken projects are lighting upgrades, increasingly using LED lamps, and the installation of solar PV. However, when these and other energy-efficiency projects are to be carried out simultaneously, they are inter-dependent and this leads to uncertainty in deciding the most appropriate scale and sizing of each type of project.

A significant block to the take up of energy performance contracts in schools and similar low energy organisations is the upfront cost, with a risk that the costs for preparing the Investment Grade Proposal will exceed the savings arising from the project payback. Reducing these upfront costs facilitates effective Energy Performance Contract development, with associated energy savings and carbon emissions reductions.

The aim was to develop a tool to examine the options and inter-dependencies of solar PV installations together with energy-efficiency schemes, principally lighting upgrades but also other electrical works, in order to derive the optimum solutions for the size of each project and deliver clients the greatest level of energy/CO_2 savings in the most cost-effective manner.

PROJECT ACHIEVEMENTS

A simple to use tool was developed using macro-enabled Excel files. The tool enables the engineer to input all relevant parameters including PV output, total loads, energy savings and all costs. By changing any of these variables the tool calculates the effect on energy and CO_2 savings and signposts the optimum solution in terms of energy savings and solar PV.

The tool was first used on a project with Cambridgeshire County Council, at Great Paxton Primary School, and

BOUYGUES ENERGIES & SERVICES Our energies for a Better Life

revealed that undertaking a lighting upgrade, with the effect of reducing the total electrical demand, would lengthen the payback for a planned solar PV project from 10.73 years to 11.29 years. On this basis it was decided to continue with the PV project as the parallel work on lighting efficiency would not have a significant impact.

The tool is flexible and can address any solar PV and lighting schemes and is now being further developed to incorporate alternative energy reduction projects such as VSD drives and voltage optimisation.

WHAT DID THE PROJECT INVOLVE DOING AND WHY DID YOU DO IT?

Bouygues Energies & Services (BYES) provides a range of energy services to clients in the private and public sectors and is a framework contractor on the REFIT programme.

Two of the most popular and commonly undertaken projects are lighting upgrades, increasingly using LED lamps, and the installation of solar PV. However, when these and other energy-efficiency projects are to be carried out simultaneously, they are inter-dependent and this leads to uncertainty in deciding the optimum scale and sizing for each part of the project. Independently designed measures also lead to overestimation of energy savings.

One of the key blockers to the take up of energy performance contracts in schools and other low energy consuming organisations is the upfront cost outlay to produce the intelligence, without the costs for undertaking the Investment Grade Proposal being more than the savings arising from the project payback.

Reducing the upfront costs through streamlining and efficiency improvements along with proactive data management allows cost effective Energy Performance Contract development, which in turn facilitates all parts of the UK to take up and benefit from Energy Performance Contracts and the associated energy savings and carbon emissions reductions.

The objective was to develop a tool to examine the options and inter-dependencies of a solar PV installation with energy-efficiency schemes, principally lighting upgrades but also other electrical works, in order to derive the optimum solutions for the size of each project and deliver clients the greatest and most accurate level of energy/CO_2 savings in the most cost-effective manner.

The tool was developed in-house using Macro-enabled Excel (.xlsm files). So costs were limited to staff time and the project took one engineer approximately three months, working around his day-to-day workload.

Solar positioning throughout the year using geometry was used in order to calculate the solar radiation levels and it took two months to complete the solar PV calculator element of the tool. It then took a further month to manipulate the lighting schedules in order to convert these into half-hourly consumption points.

The tool inputs are:

For the lighting scheme

- Existing installed system: number of fittings, light type, power, run hours
- Proposed system: proposed module with wattage, quantity, costs, controls etc.
- Run hours model: dates of operation, dates of holidays, times of operation during day, diversity factors on operational hours.

For the Project Energy Sheet (.xlsm)

- AMR (metering) Data
- Savings due to lighting (on a half-hourly basis)
- Any other instant electrical savings on half hourly basis.

For the PV calculation

- Module location (geographically)
- Module tilt (from horizontal)
- Roof size
- Module efficiency and degradation
- Shading factor
- Revised baseload (derived from the Project Energy Sheet).

The tool was initially used to look at the installation of low energy lighting alongside a solar PV array for Great Paxton Primary School, with the following results. For the PV alone the payback period was calculated as 10.73 years, with the lighting scheme include the payback for the PV scheme stretched to 11.29 years. In this case the difference was not significant (circa half a year difference), but shows the interactive effects of other Energy Conservation Measures.

WHAT ARE THE BENEFITS?

This tool is an elegant and easy to use solution to a longstanding and ongoing problem, namely how to assess the differing costs and benefits of complementary schemes in order to implement the most cost-effective solution and in such a way that the costs of preparing the desktop assessment and investment grade proposal do not outweigh the benefits.

It also allows companies to predict the energy savings to a greater level of confidence, since savings can be affected by other measures introduced to a project. The tool facilities this in a simple, flexible manner that can be used in most situations.

By the nature of Bouygues' market, the tool will be of particular benefit within the public sector and notably to the education sector, facilitating the most cost-effective savings.

ARE YOU PLANNING ANY FURTHER DEVELOPMENTS?

The next stage is to develop the tool's flexibility, to enable it to be easily used for further energy-saving schemes other than lighting, including voltage optimisation, VSD drives and chiller upgrades. There is also an intention to expand the tool further to enable it to incorporate schemes that involve gas usage.

The tool will be kept with the company, as it is part of our commercial offering, but could be used for any client portfolio.

BOWERS ELECTRICALS

POWERING A GREENER BRITAIN

PROJECT AIM

Bowers Electricals is one of the UK's biggest manufacturers of energy-saving power and distribution transformers. Proudly designing and manufacturing in Britain where our cutting-edge engineering is world renowned, our prestigious list of clients spans public and private sectors and we work on everything from power stations, government buildings, hospitals and universities to wind/solar farms and gold mines across Africa.

The business began in 1947 as an electric motor and transformer rewinder but in recent years, ours has been a tough industry to stay afloat in. More and more of our competitors now import their products from overseas, where labour and materials are cheaper. After seeing some of our competition fold and because we were in no doubt that we wanted to maintain our manufacturing roots here in Britain, we decided in 2011 to take a slightly new direction – and something of a financial risk.

We invested heavily in developing a range of transformers that focused much more on energy-saving as it was something the Bowers family felt passionately about. In 2014 we launched two innovative new products, which we believe are among the most efficient, compact and advanced in the world and now Bowers leads the way in energy efficiency.

PROJECT ACHIEVEMENTS

In 2014 we launched the Bowers Intellivolt and the Ecodesign range. The groundbreaking Intellivolt is the first 'intelligent transformer' in the world to incorporate the voltage optimisation technology of Green Apple Award winners iVolt and offers two energy-saving solutions in one package – reducing installation and maintenance costs as well as transformer losses and wasted energy.

We also unveiled the Ecodesign range, in line with the EU's new Ecodesign directive. Part of the EU's 20-20-20 aim, the legislation could result in savings of approximately 16TWh a year – or a staggering 3.7 million tonnes of CO_2 emissions – by 2020. Transformers like ours can make a real difference to the EU's energy consumption. Globally, Bowers products reduce CO_2 to the tune of 2,500 metric tonnes a year – the same absorbed by around 23 acres of forest or the equivalent of taking 520 cars off the road.

As a business, we invested an incredible amount of money, time and energy into developing products that lower the carbon footprint of our customers and reduce their operating costs.

We combine the best commercially-available materials with the most technologically-advanced manufacturing techniques to offer savings up to £115,000 over the course of a transformer's average lifespan.

WHAT DID THE PROJECT INVOLVE DOING AND WHY DID YOU DO IT?

The recession was a key time for our business. While other companies folded, we used it as an opportunity to develop new technology and products and the past two years have been among the most successful in Bowers' 68-year history.

Our project with iVolt came to fruition with the unveiling of the Intellivolt at the Sustainability Live show last April. It was our first ever exhibition and we made our debut with a very sophisticated transformer that incorporates fourth generation voltage optimisation technology.

It's been a fantastic addition to the Bowers offering not just because of the savings it promises customers, but because it firmly put Bowers on the map and introduced us to an entirely new, eco-conscious audience.

Because of the innovation we demonstrated, we won two business awards (the first we'd ever entered) and have just been nominated for a prestigious national Family Business of the Year award with PwC Later in 2014 we unveiled the Bowers Ecodesign transformer range, which was developed in order to meet new EU legislation governing energy losses from transformers which, elsewhere in the industry, have been higher than they should.

We're incredibly proud of this particular range, which was developed in conjunction with the national and international standards bodies responsible for producing the new harmonized Ecodesign transformer regulations.

It has prompted a number of orders from new customers and, with a second tier of legislation due in 2021, it has shaped how the business will develop over the coming years too.

As a result of manufacturing energy-efficient products we've become even more conscious that our manufacturing processes and the way we operate have as little impact on the environment as possible and we are certified to ISO 14001.

We have an environmental policy with regard to preventing pollution, reducing waste and to ensure that measures are implemented to protect and preserve natural habitats, flora and fauna. This is key when we are working on new sites, such as wind or solar farms.

On a day-to-day basis, we have everything from office recycling facilities to our own energy-saving transformers at work in the factory. We're also carrying out a review of water use. Unsurprisingly, we're particularly conscious about minimising electrical waste. Non-crucial equipment is turned off whenever possible and we reduce our travel emissions and costs by carrying out meetings via teleconferencing when we can and driving hybrid electric vehicles when our staff have site visits to make.

We track all of our recycling on-site, with all metals being recycled and oil reprocessed. This information is fed into monthly charts so we can monitor progress.

Even the vehicle we use for transporting our transformers to sites has been converted so that it zero emissions and, of course, because we source our materials from the UK whenever possible, our carbon footprint is kept to a minimum.

BRISTOL WATER

SPAWN TO BE WILD

PROJECT AIM

The European eel is critically endangered, which means it has the same protection as a panda. Bristol Water has a regulatory commitment under the National Environment Programme to deliver eel protection schemes under the EU Eel Regulation 2007, to open up migratory barriers and screen intakes.

Spawn to be Wild is an innovative measure which will contribute to achieve these objectives. By creating a fun, education and engagement programme we provide learning centred on the mysterious eel. The project aims to teach children about the eel and the challenges they face, as well as sessions on the water cycle, conservation, freshwater habitats and pollution.

Saving eels: youngsters are involved in the initiative

The project reaches out to a sector of society traditionally less likely to engage with the natural environment. Young people from socially challenged areas who less likely than their counterparts to take part in activities such as countryside walks, angling and birdwatching.

We believe that by enabling children from these schools to learn more about the importance of the natural world and the place we all have in it, we can help them to feel that the natural world belongs to them too. The project has included transport to ensure that no-one is excluded.

PROJECT ACHIEVEMENTS

- Creation of an excellent partnership between four stakeholders – Bristol Water, Avon Wildlife Trust, Bridgwater College and the Sustainable Eel Group
- Restocking of Blagdon Lake with the endangered European eel
- Delivery of an education programme to children around the European eel, water cycle, water efficiency, conservation and pollution
- Encouraging social equity by inviting schools from sectors less likely to visit the lakes to take part in the project
- Promoting an economic benefit to our Recreations department by delivering education around the Fishery and the lakes.

WHAT DID THE PROJECT INVOLVE DOING?

Schools are provided with a specialised elver (baby eel) tank including water tank, pump, filter and heater. The tanks are set up two weeks before the elvers are delivered.

This year the eels were donated by the Sustainable Eel Group, who are supporters of the project. Bridgwater College provide support on technical issues throughout.

An introductory session is delivered by The Avon Wildlife Trust, which is tailored to the Key Stage 2 students involved in the project. The first session includes the eel life cycle and the challenges the eel faces as it migrates. This is followed by a second education session on the water cycle, water efficiency and protection of watercourses.

The students keep the elvers for a month before they release them to the environment. The third education session is held as a field trip to the Bristol Water- owned Blagdon Lake. The field trip includes a walk around the woodland, visit to our eel trap and transfer, trout feed at our Fisheries and birdwatching, with an art activity.

The main event is the release of the elvers, and we ensure each child has an elver each to release into the lake. The project overall takes four weeks to complete.

WHY DID YOU DO IT?

The European eel is critically endangered, which means it has the same protection as a panda. Bristol Water has a regulatory commitment under the National Environment Programme to deliver eel protection schemes under the EU Eel Regulation 2007, to open up migratory barriers and screen intakes.

Spawn to be Wild is an innovative measure which will contribute to achieve these objectives. It is being delivered as part of our alternative measures, in agreement with the Environment Agency.

WHAT DID IT COST AND WHERE DID THE MONEY COME FROM?

The project costs £6,554 to deliver. The tanks and equipment incurred a one off cost of £2,400.

IF QUANTIFIABLE, WHAT DID THE PROJECT ACHIEVE IN TERMS OF SUSTAINABLE DEVELOPMENT, ECONOMY, ENVIRONMENT AND/OR EQUITY?

Sustainable development and environmental benefits

This is an education and engagement programme and we feel that environmental education and awareness underpins every part of sustainable development and environmental protection.

Education has included awareness and understanding of protected species (eels), resource efficiency (water and energy), environmental protection (catchment protection of watercourses) and an understanding of the richness of the water environment.

The education has been delivered by specialists from Avon Wildlife Trust and staff from Bristol Water, working alongside the schools' teachers to ensure that the project links with the national curriculum and learning aims and outcomes. A popular blog and website has been produced for the project with pupils and there is a new Facebook page.

Economic benefits

Bristol Water runs a recreations and amenities department, with a commercial fishery. We are always seeking new customers to visit our sites and enjoy our beautiful lakes. Engaging with school pupils about commercial angling as well as natural fish populations will help to support this business into the future.

Also, protection of drinking water sources and education on water efficiency helps to reduce the company's long-term costs on water treatment and investment required to meet an increasing demand for water.

The project also reaches out to a sector of society traditionally less likely to engage with the natural environment – namely young people from socially challenged areas who are less likely than their counterparts to take part in activities such as countryside walks, angling and birdwatching.

We believe that by enabling children from these schools to learn more about the importance of the natural world and the place we all have in it, we can help them to feel that the natural world belongs to them too. The project has included transport to ensure that no-one is excluded.

WHO AND WHAT BENEFITED?

The European eel, which is critically endangered. The children raise the elvers in the classrooms and they then get delivered safely to Blagdon Lake, avoiding obstacles and barriers, which they would have had to overcome. Pupils from inner Bristol schools and pupils from the Mendips who acted as "hosts" for their new friends from Bristol.

WHAT ARE THE LONGER TERM BENEFITS?

Education is always about long-term benefits. The main benefit in this case are an understanding of the challenges the endangered European eel faces, and that by creating a sustainable impact we can all prevent a loss in biodiversity. The project boasts many other long term benefits, including learning about the water cycle, water efficiency, greater connection with the natural environment and protection of the natural environment by understanding the links between pollution prevention and water supply.

WAS THERE ANYTHING INNOVATIVE ABOUT THE PROJECT?

This is the first time in Europe that a water company and wildlife trust have worked together to deliver an education programme using the European eel as an education tool. An Eels in Schools project was delivered by Severn & Wye Smokery but without the education and field trip aspects.

The project's innovation also exists in the partnerships which have formed to deliver it. Bristol Water already had a strong relationship with The Avon Wildlife Trust, but reached out the Sustainable Eel Group (SEG) and created a strong partnership with two parties dedicated to the protection of the European eel.

SEG introduced Bristol Water to Bridgwater College, up and coming eel specialists. Supporting each other we allowed them to initially nurture and bring on our eels, thus developing their expertise. This partnership is the foundation of the innovation this project is delivered on.

CAN OTHER ORGANISATIONS AND/OR COMMUNITIES BENEFIT FROM IMPLEMENTING YOUR METHODS?

Yes. This is a directly transferable project to any organisation working with eels and natural waters. We would be able to provide guidance on tank and eel issues we faced during the project and the EA requirements around permit in identifying a release location. It needs have good potential for migrating silver eels.

Demand from schools has been far greater than the number of participants possible, indicating that this is something which could be reproduced successfully elsewhere. Bristol Water has been approached by The Rivers Trust, Clyde River Trust in Scotland and West Country Rivers Trust, all of which want to replicate the project. We are opening up our Facebook page so children from all over the UK can engage around the project.

WHAT DID YOU LEARN FROM THE PROJECT?

Learning more about partnerships and developing new relationships between schools, a water company, college, European eel group and wildlife trust has been an excellent outcome of the project. Managing the technical issues on the new equipment have all be excellent learning outcomes.

ARE YOU PLANNING ANY FURTHER DEVELOPMENT?

The intention is to deliver the project throughout the next five years of the company's National Environment Programme.

BUCHEON CITY HALL, SOUTH KOREA

AGRICULTURAL TRANSFORMATION

PROJECT AIM

Yeowol Agricultural Park was the site of a municipal wastewater treatment plant (WWTP) that supplied city tap water for 20 years. As the site was part of a protected natural green area where land use for development had been restricted, the 52,422 square metre site was deserted and left in a poor state following the closure of the plant in 2003.

In full bloom: the city site above has been transformed

Bucheon established an action plan to facilitate the reutilisation of the former Yeowol WWTP site with the direct participation of Bucheon citizens. The city invited citizens to an open dialogue and encouraged their active participation. In the end, it was decided to transform the site into an agricultural park.

The rebirth of this former WWTP site as an agricultural park was only possible because Bucheon citizens actively participated and took ownership of the project. By taking the initiative, Bucheon citizens made a metamorphosis possible. The site was transformed from a shabby wasteland that disturbed the urban landscape into a beautiful, sustainable green space of urban renewal and revitalisation.

With the support of the municipal authorities and in mutual cooperation, 72 citizen groups and organisations are in charge of designing and operating individual activities and contents presented in the Yeowol Agricultural Park.

PROJECT ACHIEVEMENTS

The achievements of the Bucheon Yeowol Agricultural Park project can be encapsulated in three points:

- First, the park is a model case for the revitalisation of public facilities through the implementation of citizens' initiatives. An abandoned public facility was transformed into a vital green space through the concerted efforts of citizens, experts, and the municipal administration, providing a new environmental governance paradigm
- Second, the project considered the unique historical and local features of the site. Effective green spaces were created by retaining the form of facilities such as the quiescent basin, final sedimentation basin, filtration tank and recycling basin and transforming the surrounding area into green woods and leisure areas. Eco-ponds with aquatic plants and animals were created for children to enjoy and learn
- Finally, field education opportunities were created and community solidarity was restored. Children and youth can come and experience an agricultural environment and learn community values by sharing harvested crops

with marginalised community members. As a green space where nature can be enjoyed in the core of the city, the park is a wonderful recreational space for residents.

WHAT DID THE PROJECT INVOLVE DOING AND WHY DID YOU DO IT?

With its advantageous location between Seoul and Incheon, Bucheon has developed as a transport hub. Rapid economic growth, expanded transport networks and urban development have been accompanied by rapid degeneration into a sterile, grey city of 878,594 in only 53.44 square kilometres and 1.76 square kilometres of green space.

Since 1999 'Greening Bucheon' projects have been implemented to green small pieces of land. Initiated by a policy proposal from local citizens to landscape the former WWTP site as a park, Yeowol Agricultural Park transformed a grey space into a green, revitalised space.

In 2011 20 citizens formed a local citizens' committee and commissioned experts. Collaborating with the municipal government, the committee established a stepwise reutilisation plan, collected expert opinions, and performed benchmarking at the national and international levels.

Following debates and committee meetings, the municipal government of Bucheon readily embraced the idea of the citizens and established the Yeowol Agricultural Park.

IF QUANTIFIABLE, WHAT DID THE PROJECT ACHIEVE IN TERMS OF SUSTAINABLE DEVELOPMENT, ECONOMY, ENVIRONMENT AND/OR EQUITY?

In landscaping, care was taken to retain the original facilities' form. Former wastewater treatment basin sites were transformed into camping grounds, eco-ponds, urban agricultural gardens, landscape crop gardens, lotus ponds, lawn gardens, and the like.

Citizens planted and cultivated all plants and crops; the renewal project did not incur any significant costs.

Bucheon installed toilets and water supply lines for park visitors. The park is a popular place for enjoying nature and pursuing leisure activities. Programmes run by citizens since 2013 have attracted increasing interest and attention. he park has created jobs and spaces for citizens to participate and experience public agricultural education. The harvested crops have been shared among communities and distributed to needy citizens.

The following statistics illustrate the impressive level of participation by citizens in park activities:

- Urban agricultural education program: 3,000 people on 39 occasions
- Field experience program for pre-schoolers and students: 2,850 people on 38 occasions
- Eco-field classes and opportunities to experience planting and harvesting
- Community-sharing culture: 500 citizens; rice cake and winter kimchi-sharing ceremonies
- Community activities: heat-generating compost for soil conditioning
- Camping grounds: 6,367 people as of 2014.

WHO AND WHAT BENEFITED? ARE THERE LONGER-TERM BENEFITS?

Yeowol Agricultural Park has provided Bucheon citizens, including the young, opportunities to enjoy greenery in a grey urban environment and to learn the value of agriculture.

In the community agricultural gardens, families share experiences, thoughts, and feelings and visitors have field experiences. In addition to creating jobs, agricultural education programs convey agricultural skills.

The park is a model for urban agricultural development in South Korea. Its measures for facilitating a low-carbon economy, energy conservation, food mileage reduction, etc. are practicable in urban environments.

It has raised awareness of ecological issues and enabled citizens to take responsibility for their own environment in response to the global energy crisis and climate change.

Because of the park, Bucheon is now pursuing carbon-reduction strategies while assuring food safety by supplying food locally. The city is planning to spread the paradigm of the Yeowol Agricultural Park nationwide.

WAS THERE ANYTHING INNOVATIVE ABOUT THE PROJECT?

The park's innovative aspects are encapsulated in five targets and five taboos:

Five Targets:

- Organic farming based on eco-recycling
- Happiness through agricultural experience
- Enhancing community energy and communication
- Creating a popular local place for and through sharing
- Revitalizing a park based on a theme

Five Taboos:

- Ploughing
- Pesticides
- Chemical fertilisers
- Vinyl
- Pest control.

Ploughing creates ridges and furrows that harden fields and impede groundwater circulation. The agricultural gardens in the park, are not ploughed, but harrowed. The harrowed ground is not covered with vinyl but with rice-straw mats used for protecting wood from insects in winter.

Instead of artificial fertilisers, organic compost made of food waste is used. Nothing natural is discarded, everything is used. Organic farming based on eco-recycling improves soil and facilitates more intense CO_2 absorption and oxygen emission than conventional farming.

Rice-straw compost stores more carbon than chemical fertiliser by 16.9 t/ha after three years of application, equivalent to 62 tons of CO_2. Furthermore, urban farming can improve air quality and temperature and humidity control by absorbing airborne hazardous substances such as VOCs, CO, and CO_2 and emitting oxygen and moisture.

CAN OTHER ORGANISATIONS/ COMMUNITIES BENEFIT FROM IMPLEMENTING YOUR METHODS?

Bucheon Yeowol Agricultural Park is model for the renewal and preservation of idle spaces; it differs from similar projects in that it was proposed, designed, and implemented by citizens who took the initiative with the direct participation of local residents, senior citizens' welfare centres, rehabilitation welfare centres, youth centres, citizens' action groups, etc. Its operations are in the hands of the citizens and civil organisations involved. Many organisations and local governments have visited the park for benchmarking and its founders have been invited to present at national and international agricultural exhibitions. It is a representative case of innovation recognised worldwide.

Implementing the project improved cooperation and communication among citizens and enhanced environmental values. As an exemplary case of public-private partnership, the park won a Welfare Policy Best Practice award.

WHAT DID YOU LEARN FROM THE PROJECT AND ARE YOU PLANNING ANY FURTHER DEVELOPMENT?

First, we have learned that creative ideas from citizens can solve public problems, can be reflected in actual policymaking, and can be implemented.

Second, we have learned that citizens' initiative and participation can transform useless, deserted spaces into valuable local assets and environmental resources.

Finally, we have learned that civil organisations implementing public projects have the potential for self-sustaining self-governance. The Yeowol Agricultural Park is now in its third year. Future projects will provide park visitors and participants more opportunities to engage in activities that spread a community culture of sharing, including festivals based on sowing, sharing, and harvesting.

Operating year-round, the Yeowol Agricultural Park is to be run as a 'four-season themed park' with seedling planting in spring, a lotus festival and a swimming pool in summer, family healing camps in autumn, and sledding hills and rice-straw crafts in winter.

To create a beautiful natural space where humans can harmoniously coexist with nature, future plans include developing the park into an ecological park complex with bio-top functions.

Through comprehensive landscaping, the park will be naturally and ecologically connected with beautiful nearby ecological spaces, such as the million rose garden, cherry blossom garden, and azalea garden.

Bucheon Metropolitan City has now stepped up as a globally leading eco-friendly city.

Unlike industrialized cities in emerging nations, Bucheon developed from a small garden city at Boksa Village into an industrial town. It now stands as Korea's leading 'eco-friendly metropolitan city rich in culture'.

Located between Korea's capital Seoul and the adjacent Incheon city, Bucheon now serves as a traffic hub, where it has seen rapid economic growth thanks to the government-led New Town Development and transportation projects. However, 900,000 citizens currently crowd the small area that stretches to 53.4km. As a result, the green area per citizen (5.32m^2) is far below that of the legal minimum standard (6m^2). To this end, since 1999, Bucheon has kick-started the 'Greening Bucheon' project to turn underused sites into green areas.

Building on the previous project, since the 5th and 6th civil elections following 2010, Bucheon Mayor Kim Man-soo has introduced a new upgraded concept to develop Bucheon into a globally renowned eco-friendly city. Known as 'Upcycling (Upgrade + Recycling)', the new concept aims to go beyond recycling waste, and reinvent it into new products by creating new values with fresh designs'.

With the new 'Upcycling' concept, idle and abandoned spaces have not merely been remodelled, but have been given new spatial identities and historicity under Mayor Kim's leadership. For starters, the space beneath the elevated Seoul Ring Expressway was recreated as 'Haeguneul Botanical Garden'. Once a deserted area where all kinds of trash was left, it has now been transformed into a shady botanical garden that is frequented by Bucheon citizens.

Another best practice of 'Upcycling' is the 'Bucheon Yeowol Farm Park' recycling project. Formerly known as 'Yeowol Purification Plant', which ended service and closed down for 10 years, it has now been reborn and upcycled into an environmentally and citizen-friendly recreation park with urban farms and campsites. The existing facilities, including the detention pond, cleaning bed, filter basin, and recovery pond, were left as legacies of the purification plant. In addition, the green areas were reorganized into forests, resting spots, and campsites. It is also an all-season park, attracting citizens with seasonal events such as seeding (spring), banquets and swimming (summer), family camping (autumn), and sleighing and straw crafts (winter).

before	after

Before landscaping (former WWTP site)	After the opening of the Yeowol Agricultural Park

Settling basin of the Yeowol WWTP	Camping ground

Settling basin of the Yeowol WWTP	Eco-pond

Final sedimentation basin of the WWTP	Urban agricultural garden

Furthermore, Bucheon has set the standard for cultural recycling both at home and abroad, by recreating the halted Samjeong-dong incineration plant (waste treatment facility) into a space that converges and integrates culture and art. For example, the 'Spatial Indulgence' exhibition held in

July 2015 received accolades beyond expectations. Once a useless waste bunker filled with waste and seepage water, it has been transformed into a spectacular art and cultural centre and exhibition space. At present, the Bucheon Metropolitan Office aims to expand the centre to provide a 'future cultural platform' with multimedia hall, restaurant, and art performances for more citizens to enjoy exhibitions and cultural events.

Led by Mayor Kim, Bucheon Metropolitan City is evolving into an eco-friendly city with creative urban recycling projects under the concept of 'Upcycling'.

CANON EUROPE

SUSTAINABLE DEVELOPMENT

PROJECT AIM

The twin aims of this project are to improve the resource efficiency and carbon footprint of our laser toner cartridge manufacturing process, and to offer customers a convenient, sustainable and easily accessible method by which to dispose of their used cartridges.

In 1990, Canon became the first company in the world to establish a global recycling programme for toner cartridges, aiming to reduce our reliance on new natural resources and decrease or eliminate the environmental impact of these products.

Full recycle: the new plant in Japan has increased processing

Canon pursues this philosophy throughout its business, taking account of environmental impacts at all stages of the production process and offering new technologies and solutions that help our customers reduce their own environmental impacts. Starting in three countries in 1990, our recycling programme is now available globally in 24 countries.

PROJECT ACHIEVEMENTS

Since its launch in 1990, Canon's global cartridge recycling programme has created an industry standard for end-of-life cartridge management which has now become the norm. As early as 1992, Canon implemented a closed-loop system for cartridge recycling that avoids down-cycling of materials such as plastics which is very difficult to achieve and enables continuous re-use of the materials.

In addition, since 2003 we have achieved landfill-free recycling, with all parts re-used, recycled or energy recovered. During the programme's 25 years, we have expanded collection and implementation from three to 24 countries around the world, operating four dedicated recycling centres in France, Japan, China and the USA.

Operating regional recycling plants mean that Canon has been able to minimise return transport emissions while customers can easily recycle their cartridges and at no financial cost. This has enabled a high rate of resource recirculation demonstrating Canon's long-term commitment to the circular economy.

Canon

Since 1990, Canon has collected and recycled more than 344,000 tonnes of cartridges globally and saved over 502,000 tonnes of CO_2. By recovering, re-using and recycling parts from toner cartridges, Canon has reduced its need for new raw materials by 232,000 tonnes.

WHAT DID THE PROJECT INVOLVE?

From the 1980s Canon began to examine ways to reduce the environmental impact of its business operations and, following research, we developed a take-back logistics structure to facilitate the return of cartridges for recycling. This was a major breakthrough and an industry first at a time when recycling was still in its infancy and the notion of producer responsibility was not widespread among manufacturers or consumers.

After our initial success in creating the logistics structure for cartridge return, Canon invested in technologies that would allow mass recycling of cartridges. These included research and development by our Canon Inc. headquarters, to design cartridges easier to recycle, opening a recycling plant in China, and working with logistics partners to develop a cost-effective return process. Three further Canon recycling centres have since been opened in France, Japan and the USA.

WHY DID YOU DO IT?

Since 1988, Canon has pursued "Kyose" as its corporate philosophy meaning "to live and work together for the common good". In 1993, we established our environmental charter setting out our environmental assurance activities that help us contribute to a society that practices sustainable development.

Canon believes manufacturers have a responsibility to maximise resource efficiency, help protect natural resources, operate as an efficient company, and improve social and economic wellbeing. We therefore design products for re-use, anticipating the continued use of existing materials as resources for as long as possible. We see this as the most appropriate and effective method of resource management and are committed to maintaining our position as industry leader in resource efficiency maximisation. We continue to invest in new technologies so that even more materials from cartridges can be re-used and recycled in a closed loop system.

WHAT DID IT COST AND WHERE DID THE MONEY COME FROM?

The programme required a multi-million Euro investment into new technologies, resources and facilities and has been funded entirely by capital investment. We have continued to improve the recycling programme through capital and technological investment and through ongoing efficiencies within the programme and the recycling system.

In 2015, we opened a new recycling plant in Japan that now allows us to achieve 50% greater processing capacity, increases our disassembly efficiency and has improved sorting capabilities. The new recycling plant, which will be expanded to our other regions, represents the continuous evolution of Canon's recycling system.

IF QUANTIFIABLE, WHAT DID THE PROJECT ACHIEVE IN TERMS OF SUSTAINABLE DEVELOPMENT, ECONOMY, ENVIRONMENT AND/OR EQUITY?

Through recovery, re-use and recycling of parts, materials and components from used toner cartridges, Canon has reduced its need for new raw materials by 232,000 tonnes since the programme's launch in 1990. And we have collected more than 344,000 tonnes of cartridges globally for recycling, thereby saving over 502,000 tonnes of CO_2, a not-insignificant achievement. Our recycling centres also provide employment opportunities and help support local economies.

WHO AND WHAT BENEFITED?

The programme has benefited customers and other stakeholders, the environment and Canon:

- Customers can access an environmentally responsible recycling solution at no cost to them
- Environmental benefits have been achieved through the reduction of CO_2 emissions, elimination of waste and avoiding the need for new raw materials
- Canon has improved resource recirculation and encouraged the industry to do the same
- Local economies have benefited from investment and employment opportunities
- Competitors have benefited by learning from Canon's advances and implementing their own recycling programmes.

WHAT ARE THE LONGER TERM BENEFITS?

Canon's investment and benefits have already spanned 25 years. Going forward, we continue to explore more ways to recycle and re-use existing materials in new cartridge production by investing in new technologies to recycle even more materials in a closed loop system.

WAS THERE ANYTHING INNOVATIVE ABOUT THE PROJECT?

Canon was the first company to launch a global recycling programme for toner cartridges in Germany, Japan and the United States in 1990. The programme has always been free to customers and encourages responsible environmental behaviour. By researching and establishing the first recycling programme and then introducing a closed-loop system, we led the way in terms of re-using and recycling materials into new products and established what is now the industry standard for responsible management of end of life laser cartridges.

Closed-loop recycling is very difficult to achieve because it involves recycling the plastic (HIPS) from existing cartridges into new cartridges without down-cycling. As well as demonstrating market-leading expertise, the closed-loop recycling method enables Canon to continually reuse raw materials in new cartridge production thereby reducing the environmental impact of toner cartridge manufacturing.

Canon has expanded the recycling programme to 24 countries globally. By operating recycling plants in major cartridge consuming regions, Canon can recycle cartridges where they are used, eliminating the need to ship to Asia and reducing carbon emissions. In 2003 we achieved a landfill-free recycling system which ensures that no waste from this programme is sent to landfill.

CAN OTHER ORGANISATIONS BENEFIT FROM IMPLEMENTING YOUR METHODS?

- Our toner cartridge recycling programme has inspired our competitors to launch similar recycling programmes, further benefiting the environment
- Companies recycling their Canon cartridges can demonstrate their commitment to waste reduction and natural resource preservation to their stakeholders
- Communities benefit from a cleaner environment and reduced waste, and employment opportunities.

WHAT DID YOU LEARN AND ARE YOU PLANNING ANY FURTHER DEVELOPMENT?

Canon's commitment to recycling and waste reduction is long-established and anticipated market demands when it launched 25 years ago – such as the EU WEEE legislation. In 24 countries by the start of 2015, the Canon global recycling programme will be expanded where there is sufficient demand and where economically viable.

Through operating this industry-leading scheme, Canon has developed new technologies and methods to identify and implement re-use opportunities in even tough-to-recycle materials. We will continue to focus on minimising new raw materials in our products to increase environmental benefits.

CARILLION MORGAN SINDALL

A1 LEEMING TO BARTON IMPROVEMENT SCHEME

PROJECT AIM

The A1 Leeming to Barton Improvement Scheme (A1L2B) is a £380 million highway scheme, improving links between the north east of England and the rest of the country. Some 59,000 to 69,000 vehicles pass through the works on a daily basis and there are more than 90 directly affected residents, with a number of parish councils within the vicinity.

Bee friendly: wildflowers bloom in the office car park

Our aims on the scheme include promoting positive community relationships and ensure that our works enhance biodiversity. We have appointed full-time environmental advisory and community liaison teams to achieve these aims.

Key aspects of the scheme include working around nesting birds, badgers, otters, salmon and trout spawning seasons.

PROJECT ACHIEVEMENTS

In our first year in construction, we had several successes and an excellent year establishing our reputation in the local area around the project. Many of our environmental-related community projects and best practice ideas led to the project being awarded the Morgan Sindall All Star Award for Sustainability in 2014.

This award is recognition of the scheme's sustainability achievements, which are embedded into the culture of the project.

The project was also commended by CIRIA for its commitment to enhancing local biodiversity as part of the Biodiversity Innovation Group Awards in 2014. Four strong case studies were shortlisted for these awards including the creation of a pollinator corridor, tree translocation to a wildlife reserve, environmental school talks and pond dipping with a local nursery.

The project has also received other accolades for engineering excellence from CIHT and achieved 1,500,000 safe working hours during the first year of construction.

We have also received a large number of quotes and positive feedback from local parish councils and businesses. We have recently received positive feedback from the Environment Agency following the diversion of the Brough Beck.

WHAT DID THE PROJECT INVOLVE DOING AND WHY DID YOU DO IT?

The A1L2B project is a £380 million road improvement scheme funded by Highways England.

Due to its size and the works required, ensuring that the scheme promotes environmental benefits is a huge achievement.

While applying the standard environmental management principles and techniques ensures compliance with environmental legislation and company procedures, the project has gone above and beyond these expectations. We further ensure that environmental enhancements are sought and best practice is implemented.

Biodiversity

Our construction project traverses large areas of open farmland and countryside, as well as urban areas, and as a result a number of habitats and species need to be protected as a result of our works.

Through sensitive design and good environmental management during construction, we have managed to protect a number of species and measures are in place to enhance populations going forward.

Examples of best practice and biodiversity enhancements include:

- 47 bat boxes erected across the scheme (24 under licence, 23 as enhancement measures) to provide bat roosting and resting habitat in the local area. A range of boxes have been used from basic to large maternity boxes and yearly checks are undertaken to confirm bat presence. Recent inspections have confirmed many of the boxes to be used by bats, which demonstrates our commitment to protecting UK species
- New ponds have been created in a former quarry to enhance the existing large amphibian population including smooth newt and common toad. Terrestrial habitat has also been created around the ponds, reusing rubble and branches from site clearance associated with the project
- Parts of the project office car park have been turned into bee-friendly areas, with wildflowers present on the bunds. We have built insect hotels around the compound. Signage is also in place to raise awareness for site visitors and staff. Wildflower bunds have been extremely successful in attracting bees. This has been very popular with project staff during a time when decreasing bee numbers throughout the UK has been widely publicised in national media
- During a routine environmental inspection, a project ecologist noted a rare UK butterfly, the White-letter Hairstreak, at an area of woodland adjacent to the existing A1 roadway. This sighting was one of only four in all of Yorkshire in 2014 and in an attempt to enhance numbers of the species, we have amended project landscape plans to increase the planting of elm trees at this specific location. This species is found on elm trees and breeds on various species of elm; therefore increased planting of elm will hopefully lead to protecting and enhancing local populations
- Using recycled waste timber and working with a local charity for adults with learning difficulties, we have made a number of bird boxes. These have been erected in the local community at schools and a nearby church to provide nesting habitats for local birds.

Energy efficiency and resources

Where possible, we have promoted energy savings and resource efficiency through a combination of 'green' technology, common sense reuse of resources and innovative ideas, including:

- A 12,000 litre rainwater harvesting tank is in place at one of our project compounds. Grey water is used for dust suppression throughout the project rather than using mains or abstracted water
- Both project offices have been fitted out with an array of energy efficiency measures giving us a sustainable base to run the project. Measures include PIR lighting, intelligent hand dryers, non-concussive taps and double-glazed windows and the project's carbon footprint is measured and reported quarterly
- The project also uses 'green' plant and equipment, such as the Speedy Go range. We are currently trialling the world's first cordless, battery powered cut off saw, which is 100% emission free. Ecocabins are also used across the project, ensuring the energy efficiency is extended to site locations as well as main offices.

Waste

The waste hierarchy has been applied throughout the project. On site we use a range of skips to promote segregation and recycling, and we use dry mixed recycling bins in our offices and food recycling in the project canteen. We are also saving 300,000 tonnes of spoil as part of a landscaping scheme at a former quarry, saving material from going to landfill. This infill of material is covered under our EA approved Materials Management Plan and in accordance with planning permission from the local authority. We have also demonstrated good reuse of materials such as donating defective polypipe to a local wildlife charity for badger sett creation or untreated timber offcuts donated to a local charity for reuse as kindling.

Environmental design

The most effective time to make positive environmental decisions and minimise impact is at design stage. On the A1L2B project, we carefully considered environmental issues through intelligent design and forward planning.

The landscape design enhances the visual appearance of the road and ensures that noise mitigation is in place for the surrounding area. This is achieved through considered design including the road levels, earth bunding, noise barrier fencing and vegetation planting.

Furthermore, detailed design of a large culvert for a river led to a prefabricated otter ledge incorporated into the culvert sections. This allows local otter populations to travel along the river where banks were removed and was an excellent example of lessons learned from similar schemes.

Our continual improvement and positive environmental performance has been recognised during Considerate Constructors Scheme audits. We had consecutive scores of 44 out of 50 on our first two visits during 2014.

The project continues to strive for improved sustainability via objectives and targets set out in our Sustainability Leadership Plan. Our Sustainability Champions meet every two months to drive the project's sustainability agenda. This is reinforced through events such as Sustainability Week where we rolled out a number of initiatives focussing on resource use, carbon emissions and waste, to engage staff and raise sustainability awareness.

CBRE MANCHESTER ARNDALE

ENVIRONMENTAL IMPROVEMENT

Brought to book: energy savings in the Arndale Centre

PROJECT AIM

Manchester Arndale's 2015 Energy Management Plan (EMP) aimed to action key advances in the centre's energy performance. The centre's strategic vision is to save £1million in energy over five years (from the end of 2013).

Manchester Arndale aimed to approach environmental best practice with a holistic approach, tackling environmental training, awareness and practice through four key channels:

- Centre management – including partnership with EP&T Global to action energy saving systems
- Employee engagement – including Carbon Literacy and Carbon Arc training
- Occupier engagement – working with occupiers to promote sustainable best practice
- Customer awareness and education – hosting awareness raising and educative events in the shopping centre, as well as supporting city initiatives.

PROJECT ACHIEVEMENTS

The centre is currently on target to hit the £1million energy saving plan over the five years from the end of 2013.The 2014 energy expenditure saved since the commencement of the Energy Management Plan for combined gas and electricity is reported at £247,000.

The plan aimed to reduce CO_2 emissions by 6% against the baseline year (April 2013 to March 2014) through a 9% reduction in both electricity and gas in all landlord controlled areas. The centre exceeded these targets, saving 15.62% in electricity and 8.20% in gas. The total consumption saving in 2014 (first 17 months of the EMP) was 2,779,903kWh. This was equivalent to total carbon savings of 1,454tCO_2.

Employee engagement with environmental best practice has excelled, with two new training programmes in place – Carbon Literacy training and the Carbon Arc programme.

After joining the Green World Ambassadors programme, Manchester Arndale had its winning papers published in The Green Book. This prompted the team to create its own internal Green Book, which acts as a guide for employees on safety and sustainability.

WHAT DID THE PROJECT INVOLVE DOING AND WHY DID YOU DO IT?

Manchester Arndale sits in the centre of Manchester, with 1.4msqft of retail space, over 200 retailers and attracts

more than 42 million visitors annually. It approaches environmental best practice with a holistic approach, tackling environmental training, awareness and practice through four key channels:

- Centre management
- Employee engagement
- Tenant engagement
- Customer awareness and education.

Centre management

Manchester Arndale's 2015 Energy Management Plan (EMP), including the centre's partnership with EP&T Global, has led to key advances in the centre's energy performance.

The plan aimed to reduce CO_2 emissions by 6% against the baseline year (April 2013 to March 2014) through a 9% reduction in both electricity and gas in all landlord controlled areas. The centre exceeded these targets by saving 15.62% in electricity and 8.20% in gas. The total consumption saving in 2014 in the first 17 months of the EMP was 2,779,903kWh. This was equivalent to total carbon savings of 1,454tCO_2.

Manchester Arndale's strategic vision is to save £1million in energy over the five years through its EMP (from end 2013) and is currently on target. EP&T Global provide a reporting system to the centre, via a team of engineers. This service prompts the Manchester Arndale team to devise and action energy performance actions and initiatives, including:

- Installation of 130 energy monitoring and targeting systems
- Review of time settings for on/off control of plant
- Low energy lighting applications
- Thermal imaging survey of valves, fittings and exposed pipework of the high pressure hot water system – resulting in improvement to the insulation, saving 417,4020kWh, with a payback of five months
- Introduction of night cooling to the malls
- Continued focus on the building management system, auditing and adjusting settings to their optimal settings.
- Monitoring of maintenance works – changing parts and consumables to more energy efficient replacements where possible
- Alignment of operational team work tasks to reduce lighting requirements on nights.

As a result of the newly identified projects and in keeping with its energy management strategy, Manchester Arndale appointed a designated Energy Engineer to lead on sustainable initiatives and share best practice with other CBRE centres.

Employee engagement

All Manchester Arndale employees who have worked at the centre for over two years have undergone Carbon Literacy training to understand the correlation between increased CO_2 emissions and rising global temperatures. The course aims to inspire behavioural change, to reduce their CO_2 emissions.

The training covers the Energy Management Plan (EMP) and explains how to collectively help reduce the impact that Manchester Arndale has on the Environment and save money through reduced energy costs. The Manchester Carbon Literacy Project is part of a wider Greater Manchester initiative to help the city achieve its own challenging CO_2 emissions targets. The centre has played a pivotal part in raising the profile of the Manchester Carbon Literacy project, through staff training and annual three-day Carbon Active customer-facing events

The centre, working with its FM supplier Integral, launched a training programme in 2014 called Carbon Arc, a three-phase training programme to encourage staff to take ownership of energy efficiency measures inside of their control. Programme modules:

- Environmental awareness
- Savings in the home
- Saving energy in the workplace
- Marshall training
- Technical awareness training
- Technical foundation course.

An incentive scheme for reporting/implementing energy efficiency measures exists alongside the training, where employees can win £25 gift cards for successful contributions. This is key to motivating staff to prioritise energy efficiency and also reap the rewards of their commitment to environmental best practice.

After winning a Green Apple in 2013 and joining the Green World Ambassadors programme, Manchester Arndale had its winning papers published in The Green

Book. This prompted the team to create its own internal Green Book, which acts as a guide for employees on safety and sustainability.

Occupier engagement

Occupier buy-in to environmental initiatives is key to Manchester Arndale's best practice work, as their presence and actions in the centre directly impact the building's green credentials. Occupants receive an environmental guide, detailing the financial gains of reducing their environmental impacts. There are also updates on energy, waste and water performance and improve awareness of initiatives. Additionally there are two MA Green Team working groups for office and retailer occupiers, to share best practice.

Customer engagement

The fourth element to Manchester Arndale's Environmental best practice is customer engagement. With an average annual footfall of over 42 million, the centre holds powerful potential to influence visitors to make changes to their own green footprint.

Manchester Arndale hosts regular sustainability-focused initiatives in the mall to engage customers, including:
Carbon Active – The annual event in Halle Square promotes Carbon Literacy to visitors of all ages. Activity includes local and national MP and City Councillor engagement, a Carbon Classroom where local schoolchildren educate visitors on carbon literacy, and Manchester Metropolitan University's Kitchen Carbon Crimes interactive stand.

The Empty Shop – This initiative returned for its second year in 2015, to support homeless charity Mustard Tree. The "shop" located on the mall invited visitors to donate items of clothing. Each day the shop was emptied and all items of clothing have been donated to Mustard Tree, either to be sold to raise funds for the charity's projects or made accessible to those in need. Exclusive sales of higher-end donations were also held in the city to raise further funds. In total, 8,225 items of clothing were donated, equating to nearly 2.2 tones. Businesses and brands alone donated £42,000 worth of clothing, with 24 business and 14 brands involved. The event won a Purple Apple and Golden Apple in 2014, and the ICSC Foundation European and South African Community Support Award in 2015.

Dig the City – Manchester Arndale supports the city's urban gardening festival annually with green activity, such as garden installations and seed-planting for children. In 2014, its garden installation was upcycled and the plants were used to create a rooftop garden for staff on Arndale House.

Earth Hour – Manchester takes part in Earth Hour every March, by switching off all non-essential lighting in the centre and Arndale House offices. This activity is supported by social media and PR, in order to promote other Manchester residents to join the global event.

CBRE SPEKE

ESTUARY COMMERCE PARK IN SPEKE

PROJECT AIM

The Estuary Commerce Park was established in 1998 as a flagship business park for the Merseyside region. Located six miles to the south of Liverpool, the site has numerous business occupiers, ranging from office and communication centres to advanced technology research.

John Sloan of CBRE said:"The scope of the project was to clear three separate but linked waterscapes of overgrown vegetation filling the majority of the ponds."

It was recommended in a report by TEP environmental consultants report that at least 50% of the reeds were removed from the centre of the pond to help make the water body more visible and to encourage circulation within the pond along with aquatic life. Leaving a thin marginal strip of vegetation along the edges of the pond would deter access by fishermen.

As already stated, the Estuary Commerce Park was established in 1998 as a flagship business park for the Merseyside region. The external common areas of the public realm comprise roadways, extensive landscaped areas along the Estuary Boulevard and pedestrian pavements, the three balancing ponds, bus stops, external seating and lighting.

All occupiers contribute to a service charge for the upkeep and maintenance of the park. The cost of clearing the ponds was £10,750 plus VAT. This money was budgeted for in the 2014/2015 year.

PROJECT ACHIEVEMENTS

The work was undertaken between 9 and 20 February. Ground Control carefully manoeuvred a large 13 tonne long reach and small eight tonne excavator with weed bucket and dumper around each of the three waterways to protect the surrounding landscaped areas.

Clear water: the ponds in Estuary Commerce Park have been given a new lease of life

Excavated vegetation lay on the side of the ponds for 48 hours to enable any pond life to return to the water's edge, a dumper was used to remove the waste off site. To reduce the numbers of fish in the ponds and to detract trespassing on private land, we spoke to the Environmental Agency to gain advice and names of approved fisheries to lift and transport fish from the Commerce Park.

Yew tree fisheries were appointed and using a hand-held electro rod we humanely stunned and removed fish including Koi, Mirror Back and common carp up to 12 lb in weight, plus tench, rudd and even goldfish .

Yew Tree transported the fish to their separate holding tanks. Today, the ponds look fresh with open water, new signs of vegetation regrowing and water occupied by ducks and wildlife.

It was a successful exercise and no doubt is required to be repeated every three years.

WHAT DID THE PROJECT INVOLVE DOING AND WHY DID YOU DO IT?

John Sloan said: "Although the ponds were initially designed as part of the decorative landscape, a considerable amount of weed had been allowed to accumulate and agreed that it would be best to remove them all.

The park is private land and there are signs displaying "No Fishing" signs, but these are often ignored by visiting fishermen who have become a social nuisance and aid to litter in the pond.

By default the ponds over the years have become the habitat of a large variety of fish some disposed of by members of the public, even transported as eggs in the claws of herons and water birds.

The problem was what to move first – the weed or the fish and easier options than trying to remove the fishermen! It was agreed that by removing the weeds it would aid in gaining access to clear water space, and lifting the fish to be transferred elsewhere? Advice was sought from Environmental Agency fish movement section to comply with appropriate regulations obtaining permits and consignment notes.

Daniela Blomeley, of Ground Control, said: "Our project was to clear the excessive weed, debris and litter that has built up in the three ponds at Estuary Business Park, Liverpool.

"The reason for this primarily was that Canadian pond weed (Elodea Canadensis) had built up to large quantities in the waters and the common bull"rush (Typha latifolia) growing on top of the water and needed containing. The invasive Canadian pond weed was having a negative impact on the local environment as it dominated the waters at the detriment of native aquatic plants.

"Under Schedule 9 of the Wildlife and Countryside Act (1981 and amendments) under which it is an offence to plant or otherwise cause to grow Canadian pondweed in the wild, it was our duty to prevent it from spreading into any other waterways around the Liverpool area. The ponds depend on a fresh water pumping system automated to refill and connecting pipework to refill each waterspace. Any overflow feeds off to adjacent open landscape.

The weed likes nutrient-rich still waters but it also grows in flowing water so we have to prevent it spreading. It was evident the weed was blocking connecting pipework preventing the pumps from working, causing a marsh-like appearance with only a narrow water course being visible.

"As a submerged plant in a large area, the best method was to use large excavators with bucket attachments to pull the weeds by the root to discourage re-growth. This was then left on the side of the bank for 48 hours to allow any organisms to return back to the pond before it was removed to elsewhere on site to rot down naturally. A long-term preventative measure will be to use a long-handled scythe to cut the vegetation up to twice a year in the growing season.

"Since its removal we have seen a significant increase in the water level as the pumps are not clogged with weed blocking the channels transferring the water between ponds in a balanced way which is why the water level has risen. The work has been appreciated by the many occupiers and employees who can take advantage of the improved environment."

WHO AND WHAT BENEFITED?

John Sloan said: “While the work was carried out primarily for environmental reasons this project also benefited the local community that work in the offices nearby. By the ponds there are platforms for people to sit and enjoy a naturalistic view in what is a busy, industrial city.

“Removing the weeds and litter has significantly cleared the ponds and improve the aesthetics of the landscape, which in turn should hopefully encourage more businesses to the area.”

The Estuary Commerce Park is a main through route to Liverpool's John Lennon airport and has become a commuter link by Merseytravel buses. The Estuary Boulevard is a main dual carriageway with the three ponds (waterscapes) set into the wide landscaped borders which is seen and appreciated by occupiers, their employees, members of the public and international visitors to Liverpool.

CBRE as managing agents on behalf of the Estuary Commerce Park Occupiers will continue to maintain the ponds and landscaped areas to the highest standards of presentation.

CLARKS VILLAGE

WASTE MINIMISATION PROJECT

PROJECT AIM

In the last four years Clarks Village has increased its recycling rate from 56% to 87%. Consequently, around 430 tonnes of waste is now diverted from landfill annually resulting in 100 tonnes of carbon dioxide being diverted from the atmosphere.

This has saved the centre £13,000 in reduced landfill taxes and the subsequent reduction in the number of waste compactor lifts from the centre.

PROJECT ACHIEVEMENTS

Waste Minimisation General Overview:

- An increased rate of general waste to recycled waste from 56% in 2011 to 87% in 2014
- Introducing nine more recycling streams in the last four years bringing the total number of recycling streams in use at the centre to 25.

These additional recycling streams have increased total recycling by a further 131.6 tonnes per year. This includes:

- Coat hangers 4 tonnes
- PPE Bottles 3 tonnes
- Food waste tonnes
- Metal and waste electrical 5.3 tonnes
- Clothing and accessories 3 tonnes
- Polystyrene 0.24 tonnes
- Pallets 7 tonnes.

This is on top of other forms of more established recycling such as:

- Office paper 19.44 tonnes
- Poly bales 22.00 tonnes
- Cardboard 280 tonnes.

The total waste diverted from landfill is now 417 tonnes a year. This has diverted 100 tonnes of CO_2 from being emitted into the atmosphere.

WHAT DID THE PROJECT INVOLVE DOING?

Initiatives implemented to achieve these results have included:

- Introducing recycling as an action point at tenant reviews and communicating with area managers/head offices when there are issues.

The centre management team worked closely with the Cleaning team to gain their complete buy-in so that they would become the ambassadors of good waste and recycling practises and be able to inform and advise retailers when they are not recycling in the proper manner.

The cleaning team were also advised and given the authority to exert a zero tolerance approach with regard to mixed recycling bags and to not accept any rubbish that had recyclable material within it. After retailers had been informed initially of their responsibility, the focus was put back into the retailers' hands to deal with waste and recycling. if this didn't happen the cleaners would then report the offending stores to centre management who would liaise with the store managers to rectify the situation.

The same programme was introduced to support the catering teams together with silver units placed in all their service facilities, so that all food waste was separated out and then placed into clear bags and put in the silver units for collection.

Metal recycling which includes shop fit equipment has been implemented, with a metal skip available for all retailers to dispose of their unwanted metal free of charge.

Clarks Village is now working with a charity 'New Life' which collects clothing, shoes, and other accessories, odd pairs of any item also, as long as the product is not damaged beyond repair. This also including tents, suitcases and china and glass. The product collected will all go to a sorting house, where the charity will separate out and sell on what they can in aid of children's charities.

- Plastics recycling grades 1 to 6. Influencing tenants to make sure that everything from a sandwich packet to yoghurt pot can now be recycled. Every grade level of plastic can officially be recycled therefore we are influencing with regular communication and team meetings, as with all new recycling streams in making sure that the new recycling products can become as greater part of the recycling programme as much as other more basic product like cardboard.
- Raising awareness about the overall recycling collections
- Using Tenant League tables to highlight the good and the bad and create competition between stores to see who can recycle the most
- Raising tenant awareness by improving the communication process such as monthly newsletters, forum group meetings recycling league tables. Highlighting the importance and raising the opportunity through best practise.

Environmental benefits

- 417 tonnes of waste diverted from landfill
- 100 tonnes of Carbon Dioxide diverted from the atmosphere
- The knowledge that has been shared with stakeholder group's in an effort to create and help people to have greater environmental awareness
- Through clothing recycling, many charities have benefited from Clarks Village's donations of unwanted clothing and accessories, via New Life
- In 2013, there were four waste compactor pick-ups per month; this has now decreased to just one every eight weeks. Not only is this proof of the amount of waste now diverted from landfill, but the decrease in the number of collections is also resulting in CO_2 reduction from transportation.

Electrical developments

Moving on from the developments made within the waste, Clarks Village also invested £250,000 on installing 100 solar panels on the rooftops of stores to provide energy for one of the centre retail buildings. By installing the solar panels, the energy consumption within the building has decrease by 44% saving 38.9 tons of CO_2.

Cost savings

In 2014, there has been a total saving of £19,576.32 due to zero waste now going to landfill, the subsequent reduction in the number of waste compactors lifts from the centre and the installation of the solar system.

COONEY ARCHITECTS

LEITRIM INNOVATION AND GROWTH CENTRE

PROJECT AIM

Adaptation, reuse and extension of existing Leitrim County Enterprise Board offices and enterprise building located within the development envelope of Carrick-on-Shannon. The design focused on sound Passive Design principles, ensuring that natural ventilation and day lighting usage is maximised throughout to minimise the building's energy demand

The layout of the building was reconfigured to create a 'Hub' and a 'Hive'. The 'Hub' is a human scaled, naturally lit and ventilated interaction space where all members of the community would meet and generate innovative ideas.

The 'Hive' acts as a series of flexible working areas, 'Hot Desks' that facilitate impromptu meetings, collaboration, flexibility, buzz sessions and varied uses.

Making the upgrade: centre is energy efficient

PROJECT ACHIEVEMENTS

The project creates a bright, healthy and energetic environment suitable to act as an incubator for local businesses.

An integrated approach was taken with regard to the design to ensure an effective low energy/low-maintenance building was created.

This was achieved through three-storey glazed ventilation chimneys, which operated by way of a natural stack effect. The building achieved a B1 design BER rating, accounting for both new-build extension and refurbished areas of the building.

The existing building was determined to achieve an equivalent D rating. Design proposals incorporating renewable technologies (biomass boilers, photovoltaics, etc) achieved an A3 energy rating. However, these were omitted due to both capital and life cycle cost considerations.

WHAT DID THE PROJECT INVOLVE DOING?

The project involved upgrading works and extension to an existing building.

A careful study of the existing building and the site was carried out which informed the Design Team where the new extension should be placed within the site area.

The upgrading works included 'wrapping' the energy efficient, airtight new build element around the existing building and locating the extension on the brown field footings, of the partially demolished building. This has resulted in a building that is mostly south facing with its innovative natural lighting and ventilation system visible from the public road.

Materials were chosen to be robust; to achieve a balance between financial and environmental cost, locally available materials, to reduce transportation, and materials with low carbon and embodied energy.

No materials specified appear on the 'red' list and all are designed to maximise the life cycle performance (and have a cost effective initial capital cost).

At initial design stage a draft maintenance and operations plan was prepared, and this was reviewed at detailed design stage with a view to 'designing out' or minimising elements requiring high levels of maintenance.

At design stage, an integrated approach was taken with regard to the design of the building and energy performance ensuring an effective low energy/low-rnaintenance building could be achieved in practice, with little "performance gap" between the envisaged design and realised operational usage.

A key element of the brief was to achieve a building that has good acoustic performance, is comfortable, healthy and energy efficient.

Let there be light: the bright and airy interior

An acoustic survey was carried out to establish noise levels from the adjacent roadway.

The three-storey glazed ventilation chimneys provide natural fresh air to all spaces at the front of the building. The 'chimneys' operate by way of natural stack effect and acoustically controlled louvers are fitted where fresh air enters the chimneys. Building levels were established to avoid future flooding and the SUDS approach was used for surface water retention and15% of the site area was given over to broad leaf planting.

The design principals focused on Passive Design principles, a fabric first approach, ensuring that natural ventilation and daylighting usage is maximised throughout to minimise the energy demand. These fundamental principles, coupled with low energy systems eg. photocell based daylight control, condensing boiler operation and variable speed fans and pumps, combine to ensure an effective low energy design.

The straightforward approach with operations having individual controls was used to ensure actual building energy and maintenance costs were kept low. This innovative approach was recognized in the Irish Architecture Awards 2014, Highly Commended for Sustainability.

During handover demonstration days on site with client, Design Team, main contractor & specialist sub-contractors, were held.

The demonstration of the controls was recorded on DVD for our client for future reference and all demonstration manuals, maintenance manuals, labelling, Health & Safety documentation, and all relevant documentation, were completed and passed onto the client.

In addition, procedures were put in place to ensure that future maintenance personnel are familiar with the operation and maintenance procedures required to ensure the ongoing energy efficiency and maintenance of the building.

After six months we returned to review the building performance (along with the Contractor and M&E Consultants) and carried out post handover interviews with the centre manager and staff. We wanted to ensure that all elements, in particular the mechanical and electrical elements, were operating as designed.

The layout encourages all members of the community to meet and interact and generate innovative ideas. The flexible working areas, facilitates impromptu meetings, collaboration, flexibility, buzz sessions etc.; and act as an incubator for 17 local businesses.

This has led to improved communication and opportunities for collaboration among users. in addition, the selections of local material generated employment in the area.

Open days are held where local people and students are encouraged to come and visit the building and they are given tours and comparisons between how the building now works and the original building are shown. The iconography of the building has proven to be central to the learning objectives and function of the building

Mayor savings were made in building services; specifically in the areas of heating/cooling ventilation and electrical (lighting etc.).

There is a condensing LPG boiler plant (the building is remote from Natural Gas grid). Delivered Energy (BER) for Heating and HWS is 29.9 kWh/m2 (Notional Building benchmark 76.3 kWh/m2). Primary Energy for Fossil Fuel (LPG) is 32.9 kVVh/m2 (Notional Building 83.9 kWh/rn2).

No refrigerant based cooling system was used. Natural ventilation combined with effective solar shading, high performance glazing, thermal mass (used in conjunction with night ventilation purging) and daylight control (artificial lighting off reduces heat gains) ensure comfort cooling can be provided solely through natural ventilation.

Localised mechanical extract is provided (toilets etc.) with variable speed fans used throughout. Delivered Energy (BER) for cooling and auxiliary (Fans and Pumps) is 1.3kWh/m2 (Notional Building benchmark 3.4kWh/m2). Primary Energy for Electricity (cooling, auxiliary and lighting) is 54.3 kWh/m2 (Notional Building 90.8kWh/m2).

The building has achieved 95% occupancy and there are presently 17 micro-businesses in the building generating local employment and revenue. There are spaces to provide for training, self-learning, self-improvement, getting people back into the workforce. Every week eight to ten courses are held in the building.

The lifetime adaptability potential was interrogated at initial design stage. The external fabric of the building is designed to have a 60-year lifespan. All internal wall are non-structural, allowing for future layout flexibility and the building is laid out for future modules to be added.

All internal main service runs are located in the public realm circulation routes and are readily accessible for retro-fitting additional services.

The centre is part of the Nearly Zero Energy Buildings (NZEB), working as a reference for other communities, showing how a building can improve by selecting the appropriate materials and understanding the climate.

The new extension was positioned to optimise shelter and sunpath. The three ventilation and natural light giving glass 'chimneys' provide a healthy, naturally ventilated, naturally lit and quiet working environment.

By expressing the 'chimneys' on the front elevation, within the public realm, the community are encouraged to question the design and sustainable methods used.

CWMBRAN SHOPPING CENTRE

EARTH HOUR

PROJECT AIM

Cwmbran's passion to reduce their carbon footprint has and will remain their key driver in achieving not only best practice, but also reducing the impact on the environment.

Our aims over previous years have been focused on achieving that best practise. In the last year, new initiatives further demonstrated the commitment and efforts devoted to sustainable best practice through establishing innovative solutions.

The 52-acre shopping site demands an approach at site level that ensures every conceivable avenue is meticulously explored and that this practice is the driving factor that ensures sustainable options are considered at every opportunity.

At Cwmbran, our efforts are not limited to just utilities management, we are actively engaging with each stakeholder allowing us to establish both individual requirements, and added sustainable opportunities. This year our focus has been to reduce energy usage throughout site, whilst not forgetting our basic principles.

Our key principles are as follows:

- To further reduce energy consumption and carbon outputs through an in-depth analytical approach
- Sustainable procurement of materials and services benefiting the centre and the wider community
- To put reviewed waste management plans into practise and divert all feasible waste from landfill
- To explore new opportunities and build initiatives via ingenuity within each sector – travel, tourism, construction, waste management etc
- To engage with the community through events, better understand our ecological surroundings and commit to supporting local conservation efforts by funding and assisting in delivering environmental best practice.

PROJECT ACHIEVEMENTS

- Cwmbran Shopping Centre has this year formed an Energy Management Forum Group staff devoted to environmental and sustainable best practice. The group is a resource for the centre catering not only for the requirements of the business, but for incessantly reviewing procedures, usages, intended developments; ensuring that all measures taken are sustainable. The group has now been widened to the surrounding ward councillors, who will engage to heighten awareness to surrounding local communities
- Cwmbran is proud that the amount of waste segregated and recycled on site has increased annually every year to achieve 100% from landfill
- We have developed and implemented our 2014/15 Green Travel Plan through measuring the impacts on the environment resulting from travelling to and from the centre and finding alternate solutions for commuters and visitors
- Cwmbran Shopping Centre has organised and hosted sustainability awareness days for both visitors and our tenants
- The on-going implementation of our Green Procurement Plan has provided the centre with the opportunity to form a partnership with a variety of local organisations that rely on donations from our on-site reuse scheme This scheme ensures any reusable safe products are stored for the collection of vetted organisations through the local council. This scheme not only benefits the community, but ensures every feasible waste product is diverted from unnecessary disposal
- 2014-2015 has seen an impressive uplift in energy management initiatives. Identified improvements have been thoroughly investigated resulting in proposals for future projects in 2015 being included in the newly formatted Energy Management Plan.

WHO AND WHAT HAS BENEFITED?

To summarise the successful environmental performance gains over the past year, each improvement will be described by category.

1 Waste Management

Key areas were highlighted subjective to an on-site review of the waste operation at the beginning of 2014. Due to the nature of the site, some elements of the centre provided issues more challenging than that of other shopping centres.

Logistically, the Centre requires a roadworthy site vehicle for the collection of waste from the external service areas that run around the circumference of the site. Unlike other centres that rely solely on fork lift trucks only, five bins are collected at once, brought back to the compactor and emptied. Due to cost implications, the vehicle is the most efficient method of transporting waste around the centre.

The waste is then collected and brought back to one central location for compacting, baling or storage. Until mid-2013, the centre has only provided the bins/facilities for general waste, cardboard and food waste with the exception of glass recycling facilities in public locations. The remainder of the recycling figure has been that of the on-site recovery through operatives segregating the waste.

Through reviewing this system, the following changes have now been implemented:

- The construction of a 'recycling centre'
- The implementation of an additional bin for 'mixed dry recycling'
- The collection of additional waste streams that has ensured compliance with the changes to waste legislation in 2015
- The collection and recycling of paper, broachers, cans and plastic milk containers that is baled
- The collection and disposal of hazardous waste and WEEE
- New clear signage and tenant awareness campaign.

2. The recycling centre

A central area for the storage of recyclable materials was created. This eliminated the need to rely on external sources for the baling of recyclables and maximised the amount of recyclable materials segregated and captured on site. As a result, five different types of bale are now produced on site including paper, cans, bottles, poly and coat hangers. Brochures are palletised and glass is collected separately.

The yellow bin

Prior to improvements, the waste bins were categorised by colour- RED for General Waste, GREEN for Cardboard. The introduction of a YELLOW bin would provide the facility for capturing a range of recyclables that were previously only collected from a select few units and taken to an external recycling facility through the local authority.

3. Signage and awareness

"A new design for signage – *Cwmbran Shopping is committed to achieving 0% to landfill"* – has been drafted to accompany the tenant awareness campaign. Tenant co-operation has proven to be one of the more difficult tasks overall. The new steps taken have involved re-branding the centres image in regards to waste which will accompany the new bins. A back door leaflet has been produced

4. Sustainable initiatives

Procurement of materials discarded by others has benefited the centre greatly. A reuse scheme sees any unwanted office furniture, shop units or other safe reusable waste be offered to charitable organisations that may benefit from the reuse of such items.

Other resourceful initiatives have seen the waste products of developments across the site being used to back fill other areas under construction and the on-going composting of all green waste (flower baskets, foliage etc.) by the landscaping team that provides all of the composted soil necessary to sustain the centre's flowerbeds. The introduction of new 'shopping mall' dual recyclable waste bins enables segregation at source.

5. 0% to landfill

We are proud to declare that 100% of the waste produced at the centre is now diverted from landfill. All residual waste is now transported to a mechanical biological treatment site for processing into a renewable resource and the remainder rendered stable for disposal.

6. Energy management

Further reductions in energy consumptions throughout 2014-2015 (April – March) have ensured Cwmbran has retained our enviable previous successes. With an 11.08% saving of electricity, 21.85% of gas and 1.85% of water, year to date, Cwmbran's commitment to continual improvement is apparent. Previously, reductions in consumption were easier to achieve through controlling existing timers. Larger scale projects are now required for achieving further gains and this has been accomplished through an in-depth review of all feasible improvements to energy reduction. The following projects are under way:

- Pendragon House Heater Renewal – replacement of existing heaters for 550w energy saving units will achieve 27.5% saving on electricity within this building alone
- Daily monitoring and weekly update of BMS. In-depth control of energy management on site
- General Rees car park- LED lighting conversion of 630 fittings would contribute to a nine to 11% overall reduction across the Centre. The current savings of CO_2 per year stand at an impressive 70.4 tonnes and a further 125.2 tonnes will be saved per year on project completion mid-March 2015. A saving of £9,538.80 P/A has been achieved with a further saving of £12,128.40 on completion. The cost of the project is £24,456.72 with a payback period of 2.16 years
- Llewellyn car park – LED lighting conversion of the 723 fitting in the car park provides a saving of 126.9 tonnes CO_2 per annum. The current cost per annum of the 978x T8 lamps is £41,277.60. With the conversion to T8 24w LED bulbs, a saving of £17052 is achieved to bring the annual running cost down to £24,242.4. The investment for the project is £44,049.12 with a payback period of 1.81 years
- Tenant awareness has been a vital part of the overall energy reduction scheme on the centre. Through one-to-one communication, sustainability events and the reassurance of the centres commitments, support and encouragement provides the necessary motivation to relay the message to the occupants of the 170-plus units across the centre and the wider community
- Bus station – canopy lights were changed in September that has now shown us a saving of 40% energy usage across the canopy replaced 55w fittings with 38w fittings saving 250w of energy
- Llewellyn Road high level lampposts –replaced 150w light fittings with 30w LED fittings in October, producing 80% saving from 2000w to 360w
- Powys House service area – replaced 150w flood with 50w LED in November
- Prior to installation of our Christmas lights in November all lamps, frames and decorations were assessed and investment was made to purchase new fittings, which enabled us to be 100% LED lighting. Going forward, when energy usage is evaluated for the same period year on year, it will identify substantial savings.

DAVE CLARKE & SON

RENEWABLE WORKING SHOWROOM

PROJECT AIM

Our aim is to preserve the future of our own company by showing and inviting our own client base of 34 years along with prospective new clients to see just what we as a small company can achieve.

Our technologies within our wonderful old building and the passion within our showroom ensures that the clients who come to visit us from all over the United Kingdom are not disappointed by what we offer.

Our business sector and its consumers are eager for more information and clarity around the renewables market and our building is vital in providing this. We show consumers how to take control of rising energy costs by harvesting what comes naturally – daylight for one. Solar PV for harvesting electricity and solar thermal for hot water show how we can control the contamination of drinking water and air pollution.

We easily show how to look at ones carbon footprint and our aim to deliver examples of simple plumbing, right up to complex and bespoke heating systems by incorporating multiple renewable appliances is proving successful.

PROJECT ACHIEVEMENTS

We have achieved so much. In 2012 we were awarded water efficiency installer of the year for a fabulous rain water harvesting project that we installed locally to us.

In 2013 we were awarded Gold Apple Status, which in itself has taken us to some fabulous places to not only to be awarded with further awards but to see some fabulous places, projects and people.

In 2013 we were awarded energy efficiency installer of the year at the prestigious H & V awards at London's Grosvenor House hotel. In 2014 we were finalists in two categories energy efficiency and renewable project of the year. It was a fabulous achievement for us reaching the final eight out of 70 top companies in the UK. In 2014 we had lunch with the Duke of York and were asked to take part in the live pitch and panel debate for his New Entrepreneur of the Year award at St George's Hall in Liverpool. Having been asked by the Heating and Ventilation awards to be on their judging panel, we shall be at their 21st anniversary awards at London's Grosvenor in April to judge the Domestic heating and Ventilation product of the year. We feel that we are taking major strides in exhibiting and implementing what we do best and we know that our reputation continues to grow.

We exhibit many quality products in our building and recently one company came to film us to produce a video to distribute to all of their companies worldwide. This again we feel is a fabulous achievement for Dave Clarke & Son Ltd. We sponsor local football, rugby, hospital radio and history clubs, implement safety booklets into local schools and refurbished a "safe house" for the local police.

WHAT DID THE PROJECT INVOLVE AND WHY DID YOU DO IT?

We started our business in the recession in 1982 and were working from a back bedroom of a terraced house. Today, we are a husband, wife and son team who employ 25 people and three apprentices.

We purchased our former church in 1997 and with our building of great character opened our renewable energy showroom in full in 2011/12. Our project cost in the region of 87,000 which all funded by ourselves.This was particularly difficult due to having to keep the project afloat and customers and their projects happy.

In particular we look at the fact where architects and consultants would have to employ three or four companies. We do everything under one roof with a well-integrated team. To support this we have an electrical division, a renewable one and our traditional and original Dave Clarke & Son Ltd.

One of the most difficult elements was trying to overcome the mindset of customers who were not grasping the renewables message, so what do we exactly offer?

Our reception is warmed by a fabulous MCZ pellet boiler which is connected to an underfloor heating system. To show the quality of this, part can be seen through an area exposed by specialised glass. Bespoke tiling all carried out in house adorns the hand crafted reception desks and a traditional log burner merrily burns away keeping the front of house warm before we actually move into back showroom.

To the rear and in our main show area we have an impressive array of 12 appliances interlinked with a large 475 litre Xcel unit and a massive 900-litre Feuron tank. We find the heat bank perfect for bespoke hot water systems that have a lot of renewables linked together.

You can plug just about any combination of renewables into it: biomass boilers,wood burners, heat pumps and solar all work fantastically from this one unit.

We have wood burning fires, a log burning cooker which on most days produces lunch for our team, air source heat pumps, ground source and solar. We have rain water harvesting to service our facilities and our range of energy efficient lighting is demonstrated throughout the whole of the building.

WHO AND WHAT HAS BENEFITED?

We know that as a company we are taking major strides in the integration of renewable heating options and our reputation continues to grow. We can distinguish that it is about using quality products to produce quality designs which requires thought. It's about recognising each product and how they can complement each other to produce the most energy efficient heating system that you can buy.

We feel that we meet many environmental concerns by the education and encouragement of renewable energy projects. The national target for 15 per cent of the UK'S energy to come from renewable sources by 2020 and the energy technologies that we offer are now becoming increasingly important.

This all leads to climate change and issues related to global warming, the greenhouse effect and ozone depletion, we prove that we offer sustainable development aims, meeting human needs while preserving the environment.

These needs can be met not only in the present but also for generations to come. We can meet the needs of today and the challenge of tomorrow. Our showroom proves that our strength comes from its technology and the products based in it.

We were ultimately pleased to see our very own Hinckley and Bosworth Borough Council at the Green Apple awards last year with whom we have worked for many years. The council received green apple status for Hinckley's own Atkins building on which we carry out the maintenance. Again we were promoting these awards and best practice.

We sincerely hope that you will consider us for this award which will not only be extremely good for business, but will also allow us to educate further within the renewables sector, proving that we can tackle the sector's high carbon footprint and continue with the progression of promoting renewable energy.

We feel that our record is proven with sustainability as a company and hope that we can capitalise further, by reaching and developing relationships with this award and all of what the fabulous Green Apple awards brings to everyone nationally and globally.

DOLPHIN SHOPPING CENTRE, POOLE

CHEMICAL-FREE SHOPPING CENTRE

PROJECT AIM

While the primary responsibility is to its customers and retailers, the Dolphin Shopping Centre (DSC) is obligated to protect the environment and to maintain a sustainable and commercially viable shopping destination.

DSC is a large centre with intensive cleaning requirements which meant the conventional cleaning protocol involved the use of cleaning chemicals. In late 2014 the centre embarked on a project to address the number of different chemicals being used to clean the shopping centre.

The aim was not just to introduce a chemical free cleaning solution, but also to prove this would clean the centre to the standard to which it was accustomed.

Orbio is a leading innovator in sustainable technologies that provide effective cleaning and replaces multiple products with just one solution. The system uses a salt-based, chemical-free process, where water passes through the system giving it an electrical charge. The solution cleans and sanitizes leaving surfaces 99 per cent germ free. The Orbio system was trialled in late 2014.

PROJECT ACHIEVEMENTS

Prior to introducing Orbio the DSC was being cleaned every day using 14 different chemicals, from floor cleaner to lime scale remover.

Introducing the Orbio system removed chemicals from the daily cleaning regime and is also projected to save an estimated £12,000 a year on cleaning consumables. The reduction of environmental impact as compared to conventional packaged cleaning chemicals was as follows:

- Ozone depletion – 97%
- CO2 emissions – 73 %
- Acidification – 54%
- Eco-toxicity – 43%

Cleaning up: centre manager John Grinnell

The DSC is the first shopping centre to introduce the Orbio concept into its standard cleaning regime and we believe we have opened a pathway for other retail centres and high foot traffic areas to move away from chemical cleaning. We believe that not only is the DSC a more sustainable centre but it has also provided a possible positive flow-on effect to other centres by using this centre as a precedent.

WHAT DID THE PROJECT INVOLVE DOING AND WHY DID YOU DO IT?

DSC has been a historic user of traditional cleaning chemicals. That was until recently when the centre embarked on a project to move away from chemicals that contain things like phosphates and bio-accumulates that build up in the environment and are potentially harmful.

The harmful chemicals that are common pollutants are:

- Phosphoric acid
- Orthophosphoric acid
- Methanesulphuric acid
- Alkyl alcohol ethoyylate
- Ethanol
- Salicylic acid.

This is just a snapshot of some of the many chemicals used in the 14 cleaning products that were used in the centre. These have ecological effects on marine species, algae, crustacean, fish and aquatic benthic organisms (species that live on the lake or ocean bed). These cleaning products also had a sizeable environmental footprint caused by packaging, shipping and disposing of all the packaging waste.

WHO AND WHAT HAS BENEFITED?

DSC has a vested interest in the environment and its community, especially being located in close proximity to beautiful outdoors and marine life. DSC investigated a variety of chemical-free solutions in the commercial sector as part of this project with the aim to make a switch away from the status quo.

The centre identified the Tennant's Orbio os3 system as a market leader in chemical-free cleaning and arranged trials to the test technology within the shopping centre. The Orbio system is quite simple and requires minimal staff training, although initially we undertook a staff induction to the system to ensure we performed the cleaning to specification.

The consensus with the DSC cleaning team was the simplicity of the system compared to the previous cleaning regime. Not only does Orbio reduce cleaning costs in terms of products, it also simplifies the cleaning process and subsequently training time required.

As there are no chemicals we have eliminated the risk of burns and irritations sometimes seen to the skin and eyes of people in contact with cleaning solutions. Another major benefit is we leave no residue post clean as the solution is certified to be high traction by the National Floor Safety Institute (NFSI), meaning a reduced risk of slip and fall incidents.

The historic cleaning system required deliveries of chemicals once a month, which entailed a round trip for the supplier of approximately 30 miles in a 13.5 tonne vehicle. It was estimated to produce 475 grams of CO_2 per kilometre which meant the switch in materials has resulted in eliminating 0.276 tonnes of CO_2 per year.

Introducing the Orbio has advanced the DSC's sustainability efforts. The system has a very low environmental footprint, even compared to "green" cleaning chemicals, because there is virtually no packaging, shipping or packaging waste that ends up in the landfill.

WHAT DID IT COST?

Orbio is not a price restrictive technology. In comparison to the existing system the DSC will see up to 96% cost reduction compared to conventional packaged chemicals. In total, the cost for the Orbio system was £4,497.50 compared to £12,000 on the previous system.

CAN OTHER CENTRES BENEFIT?

We are championing this project and technology for other centres and high traffic areas to look into as DSC is the first shopping centre in the UK to eliminate chemicals in favour of an Orbio system, and our results speak loudly in favour of the switch.

Last year the DSC was a finalist in the Sustainable Tourism Award category at the Dorset Tourism Awards. The centre staff worked tirelessly to help raise awareness of important environmental issues and this was acknowledged at the awards ceremony.

As well as submitting information on the Orbio system, other initiatives at DSC including a Don't Drive Wednesdays campaign and a recycling and food waste scheme were put forward as part of the entry.

The DSC has employed a number of significant advances in such a short space of time and to be recognised for its commitment to sustainability, was a huge achievement.

DUBAI MUNICIPALITY

REPLACEMENT OF FLUORESCENT LIGHTING WITH LED BULBS

PROJECT AIM

As a result of the Memorandum of Understanding signed between Dubai Municipality (DM) and Philips in October 2013, Philips has conducted a study of 15 premises of the Dubai Municipality, and developed a proposal for refurbishment of the existing lighting infrastructure to an energy efficient and sustainable lighting solution.

Philips and its partner Al Ghandi Electronics executed the implementation between May and November 2014.

Leading lights: the new system is welcomed in the Dubai Municipality

The primary objectives of the project were to estimate the potential for energy savings through lighting and implement the energy savings methods.

The measure used in the project was LED luminaires of different specifications. All applied products were supplied by Philips and approved by the General Maintenance department of Dubai Municipality. To provide DM with the highest quality level of the lighting installation, Philips and Al Ghandi took responsibility of the following:

- Surveying the sites including counting of luminaires, identifying location and installation checks.
- Project management /site supervision during the installation period
- Implementation of corrective actions during and post-installation upon self-assessment as well as indication from DM
- Each building was handed over by means of a taking over certificate issued by Al Ghandi to DM, after a joint inspection was conducted.

After the mockup installation in HQ, Philips decided together with DM to reduce the brightness of the 60x60 fixtures because the increase in light levels compared to the old situation was found to be too big.

PROJECT ACHIEVEMENTS

During and after the refurbishment works, Philips has regularly interviewed users in the building to assess user comfort and customer satisfaction. In all cases the feedback was very positive.

Power consumption

Although lighting is generally responsible for 20% of the energy bill it is often neglected in energy efficiency projects. However, the savings that can be generated with energy efficient lighting are significant (over 50% power reduction), which means that over 10% of the total energy bill can be saved.

Lighting upgrade is a quick win

Lighting refurbishment is a relatively simple energy savings measure comparing to other methods, like modifications to AC or AHUs.

Straight after the new lighting is installed, energy savings are generated from day one without compromising on the comfort of people working in the environment.

WHO AND WHAT HAS BENEFITED?

Lighting is often taken for granted and only after an upgrade of the installation with a higher quality of light is it recognised what a difference can be made.
Implementation complexity

Due to the fact that part of the infrastructure is dated, the installation complexity cannot always be evaluated upfront. Detailed analysis from an experienced auditing team is the key to ensure timely completion of the project at the planned costs. Flexibility of the installation team is important to solve unforeseen complexities that emerge during the execution phase.

In order to meet the energy conservation targets of 30% (as a part of the Dubai Integrated Sustainability Strategy 2030) in the buildings of DM, the remaining buildings will have to be executed at accelerated speed.

Furthermore, Philips recommends DM to consider the addition of a BMS system to the scope. This can enable in applicable buildings to complete the energy reduction targets in one project.

To maximize energy savings across the building, the proposal is to combine lighting refurbishment with feasibility study and possibly subsequent implementation of the following energy conservation strategies:

- Upgrade to LED (savings over 50% on lighting load and additional savings on AC load)
- Lighting controls (additional savings of 30%)
- BMS (savings between around 20-30% on the AC, chillers and AHU's)
- Energy monitoring and communication of data via Green Screen (transparency of implemented energy saving measures and further awareness on the sustainability initiatives, which can further energy savings of up to 20% by encouraging responsible behaviour).

DUBAI MUNICIPALITY

SUSTAINABLE WASTE WATER MANAGEMENT PRACTICES

Saving energy: the solar landscape irrigation system is very cost effective

PROJECT AIM

The prime aim of the department is to enhance the operational efficiency of waste water management to meet the growing demand for waste water delivery and treated water supply at the highest levels of availability and reliability. It also aims to enhance the Dubai sewer infrastructure to meet the requirements of sustainable development and promote Dubai as global hub for trade, finance, tourism and green economy .

These were implemented by the department under the leadership of Eng Hassan Mohammed Makki:

- Solar landscaping irrigation system
- Implementation of environment friendly bio trickling filters
- Implementation/Integration of Scada system in Palm Jumeiraha islands(manmade islands).

The above sustainability initiatives by the department will be the first step towards achieving low carbon energy innovation in the waste water/irrigation sector in the region and the implementation of such projects will be playing a significant role in combating climate change, promoting economic competitiveness in the facility management industry in GCC, achieving energy security and environmental sustainability and promote green economy.

PROJECT ACHIEVEMENTS :

The following are the achievements of the sustainable initiatives:

Solar landscaping irrigation project

It is cost effective compared to conventional systems and in addition it has saved from the capital investment for drawing electrical cables from the grid to the pumps located remotely in highway roundabouts. Furthermore, it is more reliable and operates on minimum maintenance

Environmentally friendly

The solar energy as a source of fuel leads to clean, eco-friendly and decentralised generation of energy which saves fossil fuel, controls deforestation and prevents environmental pollution, provides greater energy security and control of access to energy.

Bio-trickling filters

Cost effective: The Bio-Trickling Filters use no chemicals and reduce the operating cost of the pumping station.The chemical scrubbers still operate as "polishing stage" (using hydrogen peroxide and sodium hydroxide) but they will be phased out. The monthly cost of chemicals in pumping stations C, E G & S was AED 11,863 in February 2011 and AED 5,431 in February 2014 – a reduction of 56%.

High efficiency

High H2S removal efficiency and the removal efficiency of the filter is 99.95% and results in reduced odour level in the pumping stations. The filters have higher efficiencies than traditional chemical scrubbers, and are capable of absorbing a much wider range of pollutants.

Scada system in Palm Jumieraha islands

The system has many operational benefits which includes the reliability and performance of the pumping stations, reduction in the operation/energy and maintenance costst.

Sustainability waste water management initiatives

The department initiated the following sustainable waste water management initiatives, which utilise solar energy and a submersible pump:

Solar landscaping irrigation project

The main aim of this solar landscaping irrigation project is to utilise the renewable solar energy and the treated water and establish a cost effective and environmentally sustainable solution by adopting a drip irrigation system to irrigate the plants and trees in the various roundabouts (traffic intersections) of the Emirate of Dubai.

The system mainly comprises of photovoltaic solar cells, a submersible solar pump and underground water tank of about 5,000 gallon capacity. A submersible pump operated by solar energy is installed in the tank, which is specially designed to run on DC current consuming 250 to 300 watts supplied by solar cells. A drip irrigation system is used to irrigate the trees and ground cover plants automatically on daily basis.

The project was implemented in about 35 roundabouts and used to irrigate the plants for surface area of about 30,000 square metres and the photovoltaic system is expected to generate about approximately 15 MWH of electricity and will cut about 6.3 metric tons of CO_2 equivalent emissions annually.

This project is being implemented in parks and in remote areas such as Hatta .
The project was recognized with H2O Mena water award and Middle East electricity award under the category of innovation/application of technology, and the facility management award for the sustainability initiative category. The project was one of the finalist in the Platt's Global Energy (USA) award and International Livcom award (UK) in the Stewardship (Green Energy Supplier) and Category and Socio-Eeconomic project categories respectively.

بلدية دبي
DUBAI MUNICIPALITY

Benefits of solar landscaping

This project is the first step towards achieving low carbon energy innovation in the waste water/irrigation sector and plays a significant role in combating climate change, promoting economic competitiveness in the facility management industry in GCC, achieving energy security and environmental sustainability and promoting green economy in the region. The project has resulted in the following benefits :

- Cost effective: it is cost effective as compared to conventional systems and it has saved money from the capital investment for drawing electrical cables from the grid to the pumps located remotely in highway roundabouts
- Reliability: it is more reliable, consistent and predictable power option and has no fuel bills as compared to conventional power system .
- Low maintenance: it operates on minimum maintenance and no refueling, making them popular for remote areas, hence the operation and maintenance is very low.
- Easy transportation: it is modular in nature and is easily transported in components and are expandable to enhance the capacity
- Environmental friendly: the solar energy as a source of fuel leads to clean, eco-friendly and decentralised generation of energy which saves the fossil fuel, controls deforestation and prevents environmental pollution.

Installation/Operation of Bio-Trickling Filters (environment friendly air pollution control system) in the main sewage pumping stations:

The filters are installed and operated in four pumping stations and are implemented as an up-gradation to the existing chemical scrubbers for the removal of odorous gases (mostly hydrogen sulfide, mercaptans and dimethyl sulphide). The gas is sucked from the pumping station wet well and the mechanical screen rooms. The technology is based on a biological reactor with counter-flow scrubber-type washer which incorporates a high surface inorganic media where the micro-organisms (autotrophic and heterotrophic bacteria will grow and will remove the pollutant compounds contained in the gas stream.

The process condition like pH, water content, the nutrient concentration and gas-liquid phase contact surface area is optimised in the reactor and manufactured from a synthetic material with a very large surface area, producing a large void volume, thereby providing a large exchange capacity between gas and liquid phase, low-pressure drop, high level of biological activity and long service life.

Benefits of bio-trickling filters:

Cost saving: the monthly cost of chemicals has been reduced by more than 60 per cent (as outlined above)

High efficiency: high H2S removal efficiency and the removal efficiency of the filter is 99.95 % and thereby reduces the odour level in and around the pumping stations. In comparison , the filters reach higher efficiencies than traditional Chemical Scrubbers, and are capable of absorbing a much wider range of pollutants.

Environmentally friendly: no wastes are generated in this system as this technology does not require the use of chemical reactive (or does require it in very low quantities whenever they are necessary). This fact minimiszes running/maintenance costs and eliminates the generation of toxic effluents that have to be treated.

Implementation of SCADA (Supervisory control and acquisition data) system – a sustainability tool for the effective management of the operation of waste water networks;

The implementation of SCADA is part of the department's strategic plan to enhance the operational efficiency of waste water networks to meet the growing demand for waste water delivery and treated water supply at the highest levels of availability and reliability.

It also aims to enhance Dubai's sewer infrastructure to meet the requirements of sustainable development and promote Dubai as a global hub for trade, finance, tourism, and green economy.

In 2013 , the department implemented integration of SCADA for seven stations (which includes about twenty pumps) located in the Palm Jumeirah islands.

Benefits of SCADA system

The system has many operational benefits which includes reliability and performance of the pumping stations, and reductions in health and safety.

The department was recognised with the Global Leadership Award in Malaysia by The Leaders International in collaboration with American Leadership Development Association (ALDA), a US-based NGO for the systematic transformation of administrative system and the waste water services delivered to public. We also won the Hariri Arab Award for Operation & Maintenance by the Arab Institute of Operation & Maintenance in Beirut under the category for the best operation system to preserve the environment in 2014.

ECO ENGINEERS AND COMPANY, SRI LANKA

NEWLY INVENTED BIOUAFF TECHNOLOGY

PROJECT AIM

BIOUAFF Technology is involved in the innovation of a new waste water treatment system that has more environmental and economic benefits to communities and industrial and commercial sectors than a currently used conventional treatment system.

Since the inception of the first research programme in 2001 we have aimed to commercialise and popularise BIOUAFF Technology globally.

Our initial research involved finding a technology for an innovative waste water treatment system which does not consume electricity during the proces and leaves no sludge. Maintenance costs are negligible and there is no odour or sound pollution. Furthermore, our system generates biogas as a by-product and this can be utilised for any fuel consumption by an organisation or institution using our system.

All the research was conducted by Mr Athula Jayamanne, the founder of Eco Engineers during his research work on a Master Degree in environmental engineering at the Asian Institute of Technology in Bangkok, Thailand.

PROJECT ACHIEVEMENTS

In Sri Lanka we successfully launched and constructed many projects in more than 20 industries including the desicccated coconut industry, hotels, rice mills, bakeries and distilleries. These industries all gained from converting to our treatment system and we have won five awards at National Level including the Presidential Award in 2009.

Since then we have received several more projects to be completed overseas. There are eight projects in India three of which have benn completed, three in Indonesia and another in the Philippines. We have also fully constructed more than 100 successful waste water treatments in Sri Lanka.

Our systems are in compliance with all environmental standards worldwide. We are constructing these systems on a commercial level in Sri Lanka and India and plan to roll them out internationally in order to cater for a global demand for convenient cost-effective and well-managed waste water treatment systems.

WHO AND WHAT HAS BENEFITED?

The Desiccated Coconut (DC) industry amounts to nearly 40% of the total export revenue from coconut products in Sri Lanka and there are approximately 40 large Desiccated Coconut Mills currently operating in the country.

Prior to 2006 there was no method of treating waste water produced from the mills in their factory process. The wastewater generated during the production of DC coconut is having a good nutrition value for biogas production, but it is harmful to the environment if released without treatment.

This is a common problem in many coconut growing countries as DC manufacturing factories emit a high-strength, low-degradable, waste water WITH a low PH value. I started to conduct research work on this regard, especially to treat the waste water coming from DC mills.

I gained my Masters Degree at AIT Bangkok, Thailand, and then returned to my native Sri Lanka where I started further research with pilot waste water treatment plants in DC mills. After conducting four years of research from 2002 to 2006, I achieved results with highly satisfactory levels that motivated me to come up with a full scale waste water treatment plant designed and built for a DC mill near Kurunegala, Sri Lanka.

Waste water produced from a DC mill has a higher pollution level including a Chemical Oxygen Demand (COD) level of 30,000 mg per litre and a Biological Oxygen Demand (BOD) level of 20,000 mg per litre (for five days). From our innovative treatment process these levels are brought down with the COD level down to 70 mg per litre and the BOD down to 20 mg per litre level.

The levels achieved by our waste water treatment process is highly effective since the COD level is almost 70% less than the required standard of Central Environmental

Authority (CEA) of Sri Lanka, the state body and authority for environmental affairs. From our waste water treatment systems a significant amount of bio gas is being produced during anaerobic filtering process (two litres of biogas is produced per litre of waste water under atmospheric pressure and temperature) and that bio gas could be used for the thermal energy requirement of the non-edible oil production of a DC mill.

Since our system doesn't consume any electricity it is a total saving to the organisation that uses our innovative waste water treatment system.

Then CEA Sri Lanka carefully studied and evaluated my premier pilot project at Kurunegala. They conducted an awareness programme for island-wide DC millers using myself as a resource person to introduce this innovative waste water treatment system, which doesn't consume electricity and that has no sludge production in the process.

The innovative waste water treatment system I introduced is an attractive, efficient. convenient, environmentally sustainable and cost effective system that could be used in non-industrial and industrial sectors of the country.

The capital cost of our system is also very low and the maintenance cost is negligible. The system is without odor and it does not cause sound pollution to surrounding environment compared to conventional systems.

On the other hand our system has the ability to generate biogas as a byproduct during the process which could be used as a fuel for a particular industry. As an example Rockland Distilleries of Sri Lanka is using biogas along with diesel – dual fuel - for their boiler and accordingly save about US $250 per day on cost of diesel as a result.

With the completion of the awareness programme, the project started to generate increasing commercial demand since many of the DC millers wanted to go for our system due to the number of environmental and economical benefits.

During this period the Coconut Development Authority (CDA) of Sri Lanka introduced a grant scheme for the DC millers to construct this innovative waste water treatment system and accordingly each miller received a maximum of up to 60% grant amounting to US $15,000 of the project cost that amounted to approximately US $25,000. This is for a plant could that treat 35 cubic metres of waste water per day.

Further soft loan schemes from the banks were also introduced to DC millers (E-friend loan with 6% interest per annum) to pursue his new waste water treatment system. Later many of our clients were entitled to another outright grant (25% of the project cost) from the Kingdom of the Netherlands. This was also available for organizations that goes for environmental control projects under a program titled Promotion of Eco Efficient Productivity Projects in Industries.

Amidst growing commercial demand we explored more potential opportunities in commercialising our system and later applied for patent rights with advisory from the Investors Commission of Sri Lanka

Our footprint has reached to various industries including hotels, condominiums, banks, hospitals, animal husbandry, rice and virgin coconut mills, garments, a bakery and distilleries. Up to now we have completed projects for over 20 hotels.

One such project was the Eden Hotel & Spa at Aluthgama that has 168 rooms where waste water generated amounts to approximately 100 cubic metres per day at 80% occupancy levels. The hotel later converted its conventional waste water treatment system (which was totally out of order) to our system with BIOUAFF Technology and now the system has been functioning smoothly for three years.

Due to haphazard conventional methods used in waste water treatment industries prior to the introduction of our system, we have found out from our research and analysis that conventional waste water treatment systems' storage pits emits a huge amount of methane into the atmosphere. But our systems has the ability to collect the methane (biogas) and burn it out as a fuel.

Now that we have understood our system, we can market it commercially around the globe as a sustainable waste water treatment system.

ECOPAINT, ANGOLA

OCTOPARK

PROJECT AIM

The new OCTOpark represents the contribution of OCTOBUILDING ® in national, European and global strategy, with a commitment to green growth and striving for principles of rational and environmental responsibility.

It intends to be a business and technological development complex, a reference to professional, cultural and recreational activities, with a strong focus on the environment and with the creation of diversified spaces.

It is always based on sustainable building solutions and the inclusion of environmentally friendly materials, recycled or recyclable. It is seeking energy and water self-sufficiency, waste treatment, resources management, in effect a reference towards best practices and environmentally responsible behaviour.

Taking advantage of the strong increase in demand for green goods and services on a global scale, and economic and employment opportunities associated with the green growth, OCTOpark follows an investment aimed at sustainable development and growth, based on strong environmental, social and economic principles.

PROJECT ACHIEVEMENTS

The OCTOpark will thus represent a nerve business centre for the region, with a significant potential for regional, national and international dissemination.

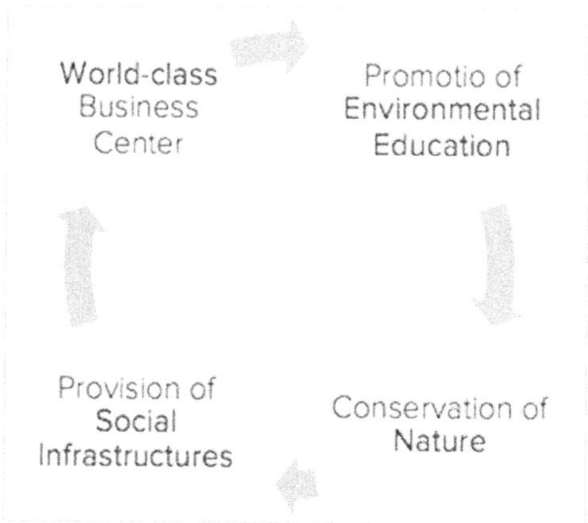

WHO AND WHAT HAS BENEFITED?

OCTOBUILDING

Briefly, the main areas and uses of the building are as follows:

- Administrative areas and headquarters for central services
- Research & development (R & D)
- Exhibitor to present sustainable solutions, and ecologic, recycled and recyclable materials
- Outdoor covered auditorium for lectures, conventions and events
- Modular offices of various types and layouts.

Recreation areas

- Open outdoor auditorium for audio-visual and outdoor events
- Leisure, recreation and sports with a gym and spa
- Small trade (food and ecological crafts)
- Astronomical observatory
- Support and social awareness areas
- Farm to show endogenous flora and fauna
- Multiple infrastructures for animals, to keep them in their habitat in a controlled and secure manner;
- Cultivation and agriculture.

Hospitality, services & catering areas

- ECO-hospitality via an organic pastries outlet and a restaurant
- Nursing home
- Playground.

Common Areas

- Pedestrian paths
- Routes for electric vehicles
- Dock stations for charging electric vehicles
- Lakes and landscaping.
- Water and waste water capture and treatment;
- Onsite capture and power generation;
- Centralized management (home automation, equipment and infrastructure) and data centre;
- Heating, ventilation and air-conditioning (HVAC);

- Waste separation and recovery (biogas production on site).

The strategic vectors, which govern and are materialized in OCTOpark, and aimed at promoting sustainable development, are:

- Reinforcement on economic activities with strong green principles
- Optimal management of resources;

FUTURE DEVELOPMENTS

We will continue to focus on the promotion of activities maximizing the protection of the environment, particularly through partnerships with centres of knowledge, such as universities and research and development centres), and with the promotion of ECO-training with companies, entrepreneurs, employees and the community.

We expect this project raises the necessary interest in green issues, focused on innovation and responsibility.

EDT AUTOMOTIVE

POLLUTION AND EMISSION REDUCTION

PROJECT AIM

We have a unique machine which we lease to motor vehicle workshops which dramatically reduces both petrol and diesel emissions and enhances fuel efficiency at the same time.

It works on vehicles as both a preventive and corrective solution. This process has never been introduced in the UK before and the general public and organisations are learning the benefits of our process. Until we formed in 2013 there was no effective way of removing engine debris which contributes to pollution.

PROJECT ACHIEVEMENTS

Since the company started, we have reduced the emissions on 13,.000 UK cars by an average of 58% on petrol engines and 66% on Diesel engines as measured in MOT test procedures.

We have many customers reporting real live savings on fuel of up 23%. No other company has our equipment: we are the sole provider in the UK.

Our number of cars treated grows every month and we have engaged both a marketing and PR company to educate the general public of the benefits of the treatment.

We now have the ability to expand into HGV, agricultural, industrial, petrol and diesel engines as soon as we can.

EIBE PLAY

SPORT ENGLAND PRIMARY SPACES

PROJECT AIM

Eibe Play is working in partnership with the Primary Spaces fund to help primary schools, who have little or no outside space, to maximise their play potential and recreational offering, so that children anywhere can enjoy safe and enjoyable new facilities.

Each eibe project has been expertly designed by our specialist creative team, so that they can be readily incorporated into the smallest of spaces regardless of location.

The new play spaces aim to not only optimise the limited acreage, but also to enhance the area and bring an aesthetically pleasing feel to what typically might be a run-down neighbourhood.

Ultimately, eibe wants to supply schools with various brand new sports facilities and multi-use games areas to encourage physical activity and improve PE lessons, so that we, as a nation, continue to nurture Great British Olympians for the rest of the 21st Century.

PROJECT ACHIEVEMENTS

Mike Diaper, Sport England Director, said: "When launching this fund, we knew there were currently many primary schools that had little or no suitable outdoor space for PE or sport, making it difficult for teachers to deliver exciting and engaging lessons.

"That is why this funding is so important because it gives young people the right places to play sport and be physically active which is crucial to their development.

"By giving children the space to take part in PE and play sport from an early age, we can help them establish a lifelong sporting habit that will also help them lead to living healthy lifestyles."

WHAT DID THE PROJECT INVOLVE DOING?

Culminating in more than £250,000 worth of revenue, bright new playgrounds and games spaces ranging from six metres x seven metres up to 18 x 10 metres are already being installed in schools around the UK, with more to come in 2015. We are thrilled with the initial response to our first few sports-inspired installations which also incorporated ancillary items such as surfacing, netting, wall mounts, floor markings and surround panels.

We were particularly excited that our launch venture was attended by Olympic gold-medal heroine Christine Ohuruogu. Such support brought fabulous publicity to the scheme and served as the ideal launch pad for further projects.

Other sites at Wonersh & Shamley Green, St Marys Frensham, St Cuthbert Mayne and Ashburnham (surfacing only) have followed hot on the heels of Woodhill School in Greenwich. There are another four scheduled for completion by Easter 2015.

WHY DID YOU DO IT?

With Sport England investing £18 million of National Lottery funding to improve PE and sport facilities in primary schools around the country, EIBE knew that the autumn of 2014 could be a busy period.

The funding for the new facilities was announced by the Prime Minister David Cameron earlier in 2014 to complement the work of the Primary PE & Sport Premium.

Children and Families Minister, Edward Timpson, said: "This multi-million pound funding will help transform PE lessons for young people by turning hundreds of concrete playgrounds into great sports facilities.

"As part of our plan to prepare young people for life in modern Britain, we want all children to get involved in sport as it can help instil confidence, discipline and determination. "Through this scheme – and more than £150 million of ring-fenced PE and Sport Premium funding which we're giving to primary schools, we are helping children develop a sporting habit for life."

Eibe are delighted to be an integral part of the whole Primary Spaces project and it was a great day when we unveiled the first of many ventures – the opening of a multi-use sports facility at Woodhill School in Greenwich, London. The equipment installed can be used both during the school day as part of the curriculum and for any clubs held out of school hours, so it is sure to be well used.

The school was awarded £30,000 of National Lottery money from the Primary Spaces fund to overhaul its playground and make it fun and contemporary.

Cllr John Fahy, Deputy Leader of the Royal Borough of Greenwich, said: "The borough would like to thank Sport England for their generous funding. The new play facilities demonstrate the Royal Borough's commitment to securing a long-lasting sporting and healthy living legacy for our residents including local schoolchildren.

"This is by upgrading current facilities and by working with our partners to create new popular, modern and accessible facilities across the whole borough to increase local participation in sport and physical activity."

Paul Redden, managing director of eibe play, said that it was the start of a playground revolution for many schools,

"Having one of Britain's most celebrated athletes present was wonderful," he said, "The children's response was a joy to behold. The Sport England team are ecstatic and the school is absolutely thrilled with the end result.

"If this is the launch-pad to get children more active, then it is an important day for Primary Spaces. Special mention should go to our installation team who worked tirelessly to complete the project on time."

EIBE PLAY

WORKING WITH DINOSAURS

PROJECT AIM:

In April 2014 eibe sales co-ordinator Stuart Evans picked up a Google Alert message that Horsham District Council were looking to build a Dinosaur Park to celebrate its local history.

Conveniently on his proverbial doorstep, Evans arranged an initial consultation with Anna Chapman, Horsham District Council's Parks & Countryside Manager. The plan was to build a children's play area full of tailored prehistoric themed equipment, which commemorated Horsham's heritage.

Eibe Play, with its environmental ethos, sustainable manufacturing process and innovative designs offered the perfect solution.

PROJECT ACHIEVEMENTS:

In June 2015 a year after the initial discussions Southwater Country Park welcomed its first visitors to Dinosaur Island. Although the official opening was not scheduled until July, the council did not want to miss the spring half-term holiday and the opportunity for children to get an early taster of the adventuresome fun that could be had.

Just seven days in and Anna Chapman is already delighted. "The park has been really popular in the first week. The equipment is very good quality," she said.

WHAT DID THE PROJECT INVOLVE DOING AND WHY DID YOU DO IT?

Dinosaur Islands a brand new adventure play area for Horsham Country Park. The previously run-down recreation space now features an array of fantastic play equipment, special effects such as an active volcano, red hot lava flows, a tar quagmire, a dinosaur swamp and a prehistoric jungle. It is a delightfully themed play space that commemorates a past age. It is sure to keep children entertained for many years in the future.

Paul Redden, eibe's managing director, believes the choice of equipment was critical in terms of appealing to a diversification of ages:

"Toddlers through to teenagers can use the park. For the older kids, there's the zip wire across lizard infested ditches and a really exciting aerial walkway along sheer slopes to look out over the country park lake. The small ones will love crossing the dinosaur bones and deadly swamp land. Even the parents will enjoy swinging through the jungle and climbing the volcano!"

The important specification of products stemmed from consultations with local schools during the tender process, once the shortlist had been decided. The key stakeholders in the Eastern Shires Procurement Organisation (ESPO) tender then deliberated, and taking on board the feedback from the children, opted for eibe's inventive proposal.

It was our uniqueness of products that stole the show, mixed with empathy in mirroring the council's own mission to regenerate the area and theme it in a natural way, thus creating a destination park that would ultimately entertain and educate. The initial concept was a 3D Masterpiece created by design director David Hibberd and his team. So began the detailed design stage. Many tweaks and adjustments were made to the above first draft.

"Working in harmony with Deborah Mallard, the council's landscape architect, we had to take into account the contours and landscaping of the park," says Hibberd,

“Being mindful of the standards of foundations needed for our high duty equipment was a design imperative. Several visits were necessary to plot the slopes, paths and natural swales. Back in the studio, we adapted our designs accordingly. Making a detailed video presentation for the client was a gem of an idea and brought the whole concept to life."

Installation work began in earnest around Christmas and our hardy project team met with typically seasonal disruption. "At one stage, out installers were knee-deep in clay," said project manager Ben Neaves, "The inclement weather caused some inevitable delays, but the team were determined to complete the project on time, so worked through extraordinary conditions to fulfil the promise."

Monster fun: the Dinosaur Park has become a hit with people of all ages

Anna Chapman was relieved and grateful for eibe's dedication to the cause and paid special tribute to our suppliers, RJ Playground Services. She said: "The installation team's service was really good. Indeed, so was the helpful contact from the entire eibe team."

Spring arrived and Steve Delahunt, the Senior Park Ranger, was thrilled to see the new park unveiled. The £140,000 play space was now ready for excited arrivals.

All that remained to be finalised was the surfacing and grass matting. Featuring a myriad of themed wooden robinia equipment, all manufactured from FSC timber, Dinosaur Island hosts a paradiso predator's trap, dinosaur slides, balancing trails, nest swing, toddler and tyre swings, a rope tree walk, stand-up seesaw, a sand construction unit, an obstacle course and the all-important dinosaur sandpit featuring that iguanodon fossil.

The park was completed in May and was officially opened on July 11. Paul Redden summed up the venture: "It's always more rewarding when a scheme has some sort of historical significance for us. While we have never previously gone this far back in time, we have, in the last couple of years, undertaken themed projects relating to world wars the Titanic, Fort Royal and the Battle of Hastings. Dinosaur Island will deliver fun and education to children of all ages for many years to come.

"Partnering with Horsham District Council has been a rewarding journey. Our Green World Ambassador status was certainly a core credential in us winning the tender; the client wanted someone with expertise in natural environments," added Paul.

"Our sustainability ethos is at the core of our own Corporate Social Responsibility pledge. All of the wooden equipment used at Horsham Country Park is, as ever, FSC certified. We always consider matters of sustainability in relation to the production and processing of materials — the amount of timber harvested never exceeds the amount of timber regrown.

"As a business, we will continue to focus on the area of recycling management; the crucial issue of sustainable forestry is strictly supervised.

"As we strive for continuous improvement, we conduct regular audits, use environmentally friendly design and recyclable packaging for our products, all of which falls in line with the RESY regulations

"Moving forward, we hope to work with more organisations that share our vision. It has been uplifting to watch Dinosaur Island come to fruition, a project which absolutely embraces all that is important about the past, present and future of our world."

ENTERPRISE-RENT-A-CAR

ENTERPRISE RENT-A-CAR 20TH BIRTHDAY ROUNDABOUTS

PROJECT AIM

To celebrate its 20th birthday in the UK, Enterprise Rent-A-Car wanted a regional initiative to bring lasting value to communities.

As a regional business with a network of over 400 branches across the UK, communities are vital to Enterprise's success. Sustainability and community involvement are core company values – part of what Enterprise terms its Cultural Compass, a set of principles which drives corporate strategy across the world.

The aim of the project was to give something back to communities and to deliver long-term environmental impact to everyone living in the area. The landscaping of roundabouts across the country with sustainable perennial wildflower meadows was chosen because of:

- Its link to transport, a natural fit with the business
- The opportunity to beautify areas which are often left uncultivated
- A clear need to provide a solution which would lower maintenance costs for local authorities.

Enterprise is a long time investor in ecological projects. In the US, the company celebrated its 50th birthday with a 50-year commitment to plant 50 million trees to replenish forestry and not be felled for timber. Enterprise also funds green fuel and transport research programmes and is at the vanguard of sustainable business practices.

PROJECT ACHIEVEMENTS

Partnering with the University of Sheffield Department of Landscape, Pictorial Meadows and Local Authorities, more than three acres of roundabout land will be planted with beautiful naturalistic planting schemes designed to increase sustainability, lower maintenance costs and promote biodiversity.

The project has national impact. The location of the seven roundabouts to be re-landscaped are:

- Sherlock Street, Birmingham
- Drumry Road, Glasgow

Saying it with flowers: an enterprising way to celebrate a 20th birthday

- Middleton, Leeds
- East Prescott Road, Liverpool
- Park Square, Sheffield
- Elgin Drive and Kembrey Park, Swindon
- Robin Hood Road, Woking.

In addition, seed blends and labour were provided to Elmbridge Council in order to create annual wildflower meadows across the borough. Two community planting days engaged local schools and community groups in Elmbridge.

The project has enabled the University of Sheffield and Pictorial Meadows to roll out their wildflower meadow technology to a national audience.

Professor Nigel Dunnett has spent two decades developing designs, techniques and processes to achieve consistently reliable and long-flowering performance from new urban landscape planting. He founded the company, Pictorial Meadows, to bring this technology to sites around the UK at an affordable cost, through a range of meadow seed mixes. This is one of the largest perennial meadow projects which both organisations have been jointly involved in.

WHAT DID THE PROJECT INVOLVE DOING AND WHY DID YOU DO IT?

Enterprise partnered with local authorities across the country to landscape a roundabout with the latest sustainable perennial meadow technology.

The company commissioned Nigel Dunnett, Professor of Planting Design and Vegetation Technology at The University of Sheffield, to design each roundabouts. Professor Dunnett is well-known for innovative and pioneering work in urban landscape design and planting. He was lead planting designer and horticultural consultant for the 2012 Olympic Park, together with his colleague James Hitchmough, creating stunning meadow displays and beautiful meadow landscapes for the largest new urban park to be made in the UK, and is a Gold Medal winner at the RHS Chelsea Flower Show. Pictorial Meadows, originally a company of the University of Sheffield is now the UK's leading meadows specialist, with a reputation for innovation in delivering projects on the ground.

This ecological initiative celebrates Enterprise's 20 years in the UK with a commemoration that will have lasting impact. As a regional business with over 400 branches located within 10 miles of over 90 per cent of everyone in the UK, the company wanted to give back to the communities that have helped to make it successful. There will be no branding on the roundabouts.

WHAT DID IT COST?

The £130,000 programme will be fully funded by the UK business and the Enterprise Holdings Foundation, the philanthropic arm of the company.

WHO AND WHAT HAS BENEFITED?

The planting mixes and seed blends involved in the project will promote biodiversity. In addition to bringing a blast of nature into Britain's urban areas, the programme also supports the Government's National Pollinator Strategy, launched by DEFRA in November 2014, which aims to improve conditions for bees and other pollinating insects.

It is anticipated that the designs will reduce maintenance costs to the local authorities as the planting schemes selected will require only an annual "cut and collect" as opposed to monthly maintenance of bedding plants and grass turf. These costs vary per authority but Pictorial Meadows estimates that it can typically reduce maintenance costs by over 50 per cent.

Local residents, motorists and visitors to each of these eight locations will benefit from a splash of brightly coloured nature in busy road intersections.

The schemes promote biodiversity as well as the Government's National Pollinator Strategy, and they reduce maintenance costs and benefit local authority budgets.

Councillor Brian Ford, Swindon Borough Council cabinet member for Streetsmart, said: "We've experimented with wildflower meadows in the past, so we were very interested in the prospect of working with the recognised experts in this field to improve on what has already been done. We strongly believe in the importance of sustainable planting as a way of bringing urban areas to life and we felt this project is a creative way to improve the aesthetics of the area for the benefit of Swindon residents."

WHAT ARE THE LONG-TERM BENEFITS?

The designs created by Nigel Dunnett are designed to provide stunning visual effects and ecological benefits, as well as reduced maintenance costs, for many years. In fact, several of the councils involved with the Enterprise meadow programme are using this project as a sponsored pilot for how this technology can be rolled out within their region in terms of time scales, costs and positive impacts.

This programme is one of the first privately funded projects to bring the technology into action in communities across the UK. As such, it is an important project for the Department of Landscape because it gives nation wide examples of its distinctive urban greening approach and for the research team at Pictorial Meadows because it will provide very valuable test cases for these new technologies.

Pictorial Meadows itself is largely owned by the social enterprise Green Estate who use all profit to further research, manage local parks and green spaces and share and encourage other community groups to see the Social Value Act implemented in their area.

WAS THERE ANYTHING INNOVATIVE ABOUT THE PROJECT?

Professor Dunnett has spent two decades developing designs, techniques and processes to achieve consistently reliable and long-flowering performance from new urban landscape planting. He founded the company, Pictorial Meadows, to bring this technology to sites around the UK at an affordable cost, through a range of meadow seed mixes.

The Enterprise roundabout project will combine a whole range of techniques and methods to achieve striking transformations.

Robust perennial plants, vigorous shrubs and colourful bulbs are combined with seed mixes of colourful annuals and perennials, specially selected for the climate and growing conditions prevalent in each location. The planting approach has been developed at University of Sheffield as a low-cost, simple, but highly dramatic and long-flowering way of planting urban landscapes. This is the first time that it will be rolled out nationally.

A special innovation is the cultivation of two acres of assorted 'pre-grown' meadow mats at a site in Lincolnshire to provide sufficient meadow for the Enterprise project. The mats are cultivated off-site to allow for intensive irrigation essential to ensuring the young plants establish a strong root base.

Following careful preparation of the ground, the plants, seeds and meadow mats will be put in place in the autumn of 2015.

CAN OTHER COMMUNITIES BENEFIT?

As one of the first national rollouts of Professor Dunnett's research, this project will be a test case for how other authorities and land managers can implement perennial meadow technologies on their properties.

Several of the councils involved are using this project as a sponsored pilot for how this technology can be rolled out within their region. Once completed, urban greening experts will be able to use the results of this project to test the viability of the Department of Landscape technology for their own meadow programmes.

WHAT DID YOU LEARN FROM THE PROJECT?

The timescale has been greater than at first anticipated. Local authorities have been highly enthusiastic about the project. However, due to the number of stakeholders involved, the complexity of the design and the timetable imposed by planting seasons, the full canopy of each meadow will only be in bloom in spring/summer 2016.

ENVIRONMENTAL CROP MANAGEMENT

BIOPESTICIDES AND INTEGRATED CROP MANAGEMENT

PROJECT AIM

The aim of the project is to research, develop and incorporate biopesticides into the every day agronomy of our farmers' crops.

Biopesticides are pesticides derived from natural materials such as animals, plants , bacteria and certain minerals. As agronomists(crop doctors) we are looking for the safest (to the farmer, consumer and environment) and most effective ways of controlling crop pests and diseases.

Biopesticides offer better water protection, minimal crop residues and ensure better farm worker safety. In addition they are effective in managing pesticide resistance and maintain beneficial insect populations .

PROJECT ACHIEVEMENTS

ECM has pioneered the use of mustard to control many crop pests and weeds and more recently has undertaken research at Manchester Metropolitan University on biopesticides. This proved claims that bio-pesticides could match or even surpass conventional pesticides.

ECM agronomists have used a garlic-based granular insecticide to replace pest control in carrots and potatoes and has replaced 10% of the usual anticholinesterase carbamate insecticide. As this material has a long harvest interval we have been able to protect potatoes that would have gone untreated for potato cyst nematode, a potential crop destroyer, by substituting the garlic extract.

ECM has used a ferric phosphate pellet with a zero harvest interval, no water course risk and very low impact on birds.Over 20% of conventional pellets have been replaced. Crops have been fully protected at no extra cost and our farmers are keen to increase use of these greener products next year.

WHAT DID THE PROJECT INVOLVE DOING AND WHY DID YOU DO IT?

The project involved taking previous research and experience in organic farming techniques and convincing our farmers these novel would work and protect their crops.

As many pesticides are withdrawn or restricted farmers need an alternative and complimentary technology to maintain our food production,whilst protecting water and wildlife.

Our previous experience with mustards has convinced ECM that pests and diseases can be controlled using natural agents. In fact we have seen direct evidence during our crop inspections that the immune system of the crop is triggered into protecting itself from disease and pest attack.

On our regular farm visits we have seen the farmer accept that these new materials are as good if not better than conventional chemicals and the farmers are happy with the greater safety that they offer.

Farmers' worries over the loss of many pesticides is relieved and ECM has been approached by many biopesticide manufacturers to help in the commercialization of this brand new technology.

WHO AND WHAT HAS BENEFITED FROM THE PROJECT?

- Farmer is handling safe product that is non toxic to him, watercourses and wildlife
- Effective crop protection with effective management of pesticide resistance
- Consumer has healthy pest free food with no pesticide residue
- Beneficial insects are maintained and mammals , birds and fish are kept safe.

- Farmers can continue to feed a growing population in spite of losing some essential pesticides that threatened food production in the UK.

ECM will adopt biopesticides on all our 1,200 farmers in 2015 following this successful pilot scheme. ECM has shared this work with over 90 Crest agronomists across the UK who will pass on this new approach to their farmer clients.

ECM is working with our university research team and our multinational suppliers to get fast and efficient exploitation of this green and sustainable technology.

ARE YOU PLANNING ANY FURTHER DEVELOPMENTS?

ECM is examining several biopesticides that will be marketed in the coming years. Further research on supporting conventional crop protection will also be carried out.

FLOORBRITE IN ASSOCIATION WITH CBRE AND TfGM

GREEN TRAVEL INITIATIVE

PROJECT AIM

The aim was to promote viable and sustainable active alternative green travel options for the occupants of a multi-tenanted office development in Bolton. To improve health, wellbeing, access to employment and reduce carbon. To build a green travel partnership with Transport for Greater Manchester (TfGM) and for the Building Manager to become a travel champion.

This would enable the occupiers to be better informed and incentivised to make a positive change in the way they travel to work. We would access funds to refurbish the cycle hub and promote cycling as a green travel alternative.

PROJECT ACHIEVEMENTS

In partnership with TfGM and their travel partners, we conducted an online travel survey and based on these results developed a building travel plan with objectives.

More than 70 personal travel plans were given to tenants so they could make alternative green travel decisions. The Building Manager attended numerous green travel workshops with TfGM so he could discuss alternative travel options with tenants. Discounted and corporate bus and rail travel have been discussed with tenants. The building management has organised a cycle to work week for all the tenants with free cycle maintenance workshop. The tenants have taken part in the Walk to Work Week in May 15. A successful bid has been submitted to refurbish the cycle hub and promote cycling through specific events and maintenance workshops.

WHAT DID THE PROJECT INVOLVE DOING?

Working with TfGM and their travel partners to provide over 70 personalised travel plans. Attending TfGM travel workshops so tenants could receive regular updates at meetings. Developing a travel plan with objectives for the building. Organised cycle to work events, maintenance workshops and other green travel events. Participated in National events such as Living Streets Walk to Work week and National Cycle Week. Incentivised participation by giving tenants pedometers, umbrellas, poncho pods, water bottles and cycle safety equipment.

WHY DID YOU DO IT?

The building is a multi-tenanted office in the centre of Bolton with more than700 employees from five different companies and spread over seven floors. At the start of the project there was no tenant participation in green travel events or promoting green alternatives. There was no joined up approach for green travel and options for the tenants. The cycle hub was very limited and did not promote cycling as a viable green travel option.

WHAT DID IT COST?

The only cost was the successful bid to refurbish the cycle hub and promote cycling.

WHO AND WHAT HAS BENEFITED?

Since we have started the project we have issued 70-plus personal travel plans, with more than 20 tenants stating they have changed their travel plans and used public transport. Cycling has increased seven-fold and this has been achieved by holding free maintenance events and cycle to work events.

Tenants are taking part in the CBRE Environmental Green Travel Competition and have clocked up 1,410 miles resulting in saving 78Kg CO_2 and burning 40,111 calories, saving £577. During the cycle to work and maintenance events participants have been given high vis vests, reflective items, water bottles and cycle maps for the Greater Manchester area.

The building has successfully been awarded funding to refurbish the cycle hub and it will go from having just bike racks to having a maintenance area, seating, lockers and improved storage facilities.

Tenants took part in the Living Streets Walk to Work and all those that took part were given a pedometer, poncho pod, water bottle and umbrella. Last year no tenant took part in the event whereas more than ten people walked to work this year and recorded their mileage on the Living Street website.

More than 70 personal travel plans have been issued and information regarding discounted rail and bus travel given to the directors of each company. A successful bid for funding from TfGM has been secured to refurbish the cycle area. As a Building Manager I have organised a presentation by TfGM for all CBRE Building Managers in Greater Manchester so they can advertise green travel options and future events.

Every tenant has had the option to participate in the organised events and all those who have participated have saved money, improved their health and wellbeing and promoted a sense of community within the building.

Employers have had the opportunity to access discounted bus and rail travel for all their employees. The carbon footprint of the building has been reduced.

WHAT ARE THE LONG TERM BENEFITS?

Health, wellbeing, financial, environmental, community, with reduced employee sickness absences and building togetherness which will lead to increased participation in future events. The project is being organised in partnership with TfGM and we are working with multiple partnerss/tenants trying to promote green travel options and provide advice.

CAN OTHER ORGANISATIONS BENEFIT FROM IMPLEMENTING?

If all organisations within Greater Manchester accessed the TfGM travel team, they can access the relevant grants, travel workshops and events. As a Building Manager I have organised a presentation by TfGM for all CBRE Building Managers in Greater Manchester so they can advertise green travel options and promote sustainable travel.

WHAT DID YOU LEARN?

I have learnt that tenants will participate in events and embrace green travel if they have the information and the benefits they will get from participating. Employers will also support green travel and events when they understand the benefits. We will continue to promote green travel and initiatives within Greater Manchester and build on the increased participation since starting the project.

FSC BLENCATHRA CENTRE

RENEWABLE ENERGY PROJECT

PROJECT AIM

Green outlook: the centre has a renewal energy project

The Blencathra Centre has a national reputation for fieldwork and was established by the Field Studies Council in 1993. The centre's main building is a former tuberculosis sanatorium which was converted into a residential field centre and occupies an inspiring site 300m on the south facing slope of Blencathra.

The centre comprises 12 separate buildings, mainly of traditional stone construction that have been converted to teaching and accommodation areas. With increasing costs of oil-fired heating and electricity, and environmental awareness key to our business, we developed a Renewable Energy Project informed by feasibility studies and assessments, which has radically changed our energy sources and usage.

PROJECT ACHIEVEMENTS

The Renewable Energy - Low Carbon Vision Project reduced the carbon emissions of the Blencathra Centre by 80% through the installation of biomass district heating and micro-hydropower alongside improving the energy efficiency of the existing buildings.

The project created an opportunity for the Blencathra Centre to become a low carbon exemplar tackling climate change within the Lake District National Park.

We are not aware of a successful project of this scale and complexity that has combined so many energy saving elements in one scheme. From the outset the project was one of only a handful that was successful in securing funds through the Rural Carbon Challenge Fund. This fund supported projects which could demonstrate new achievements in technology, training and carbon savings in rural locations that would otherwise be too challenging to proceed.

The hydropower scheme is a unique example. The project overcame the challenges of obtaining landowner permission, some ten statutory consents, completing construction in almost inaccessible areas in all weathers, reusing elements of a historic scheme and developing a solution to carry the generated electricity 2.5km to the grid connection at the centre.

When the project started the biomass district heating scheme was the first of this scale in Cumbria and, with 12 buildings connected to the network. Achievements included:

- 35kW hydro-electricity turbine using an existing weir on Roughen Gill connected to the mains electricity supply 2.5km away at the centre
- 300kW biomass district heating system
- Retrofitting heating systems and insulation.
- Training and educational programmers
- CO_2 displaced (214.8tpa).

WHAT DID THE PROJECT INVOLVE DOING?

The renewable project was based on sustainability, in particular the reduction of greenhouse gases and its relationship to the climate change agenda. The design and operation of each element was developed to meet or exceed the environmental standards and requirements of the LDNPA, the Environment Agency and Natural England.

The carbon saving of the project has been audited independently and concluded that 214.8tpa CO_2 have been saved. The audit also identified that the energy use of the site had increased. However, the project still managed to reduce carbon emissions by 80%.

WHY DID YOU DO IT?

The Field Studies Council is an environmental education charity and it is important at Blencathra that we try to "practice what we preach". Every year we host more than 8,000 learners from over 300 schools and universities. Our visitors come from all over the UK, and for many it is their first visit to the Lake District. During their courses, students come into contact with our approach to green tourism. It is a unique experience, enriched through an awareness of the special relationship between people and the environment.

WHAT DID IT COST AND WHERE DID THE MONEY COME FROM?

Blencathra FSC was not able to achieve a project of this scale with traditional funding sources. In 2010 we secured £500k from the Rural Carbon Challenge Fund, with £281k match funding. Following the Government Spending Review, funds were reduced and delayed until the project finally started in 2011. The total was £800k when completed in 2014.

WHO AND WHAT HAS BENEFITED?

The Blencathra Centre employs 28 staff and has a turnover of £1million and is a significant contributor to the local economy.

The training stage of the project has improved understanding of low carbon technologies and processes across the north west through the delivery of educational and training programmes .Blencathra has exceeded the targets set:

- Micro-generation businesses supported/created 5
- Jobs created 2/Safeguarded 24
- CO_2 displaced 214.8tpa
- Training days delivered 973
- Participants completing training day 15,000
- Training for micro businesses/organisations 150.

We have also provided workshops/site visits for several local organisations and businesses for example CREA, Eden District Council, Nurture Lakeland and CaFS.

WHAT ARE THE LONGER TERM BENEFITS?

Our plan is that carbon management within FSC Blencathra will be the responsibility of all staff and visitors, with a focus on behavioural change. We also aim to improve and develop how we monitor, record and communicate our energy data usage. Many initiatives have become embedded in our usual routines, for example the provision of recycling bins or the use of environmental cleaning products. However, the challenge is to ensure that all staff are aware of these achievements and improvements, to keep this fresh and develop new initiatives to keep pace with current expectations and trends.

WAS THERE ANYTHING INNOVATIVE ABOUT THE PROJECT?

A project of this scale and complexity required innovation in several areas:

Hydropower

The hydropower scheme faced several challenges and is a unique example. The fact that an historic scheme in such a location was feasible was too good an opportunity to ignore. The site is in a national park, a SAC and SSSI, on a private estate, on common land and a public bridleway. Ten formal permissions were obtained, all with their own requirements, negotiations and timeframes.

The engineering challenge to build a hydropower scheme in such a remote location was significant and required inventive solutions for access and construction. The turbine is 2.6km from the centre and the innovative solution, developed by Border Hydro, was a pair of transformers to raise and lower the voltage at each end.

Biomass

When the project started it was the first district heating scheme of this scale in Cumbria. Local company Barden Energy was able to design and build the scheme, as well as supply locally sourced woodchip.

Education

By installing these technologies we have expanded our education programmes to cover on-site demonstration of renewable energy. Blencathra has been at the vanguard of trialling and piloting rural broadband and wi-fi initially with 4G and recently with the installation of a fibre system 30Mb to support learning which has increased the capacity to deliver to business.

CAN OTHER ORGANISATIONS BENEFIT FROM IMPLEMENTING YOUR METHODS?

Switching our heating from a fossil-fuel system was crucial to reducing carbon. A range of ideas was investigated, including solar thermal and air and ground-source heating. Geology and weather considerations led us to a biomass solution. Now a 300kW biomass boiler provides heating and hot water for over 120 resident guests per night via an extensive district heating scheme.

The HEP uses the existing weir and has a 35kW turbine and turbine house. The innovation is applicable to any remote small scale electricity generation source. It can enable the energy sector to continue to develop renewable energy potential in areas that would otherwise be prohibitively expensive, inaccessible or at an unacceptable environmental cost.

We are not aware of a successful project of this scale and complexity that has combined so many energy saving elements in one scheme. The Fund supported projects which could demonstrate new achievements in technology, training and carbon savings in rural locations that would otherwise be too challenging to proceed.

WHAT DID YOU LEARN AND ARE YOU PLANNING ANY FURTHER DEVELOPMENT?

Installation of 10kW PV panels on one of the larger south facing classroom buildings is planned as part of an energy retrofit of the building. including a new roof and major structural works – when the presence of bats allows work to take place.

We abstract water from a captive spring and operate our own sewage system. It is important to consider both these together in the context of water management. An upgrade of our sewage system in 2015 should provide further interpretation opportunities.

In 2015 we have just installed an electrical charging point to capitalise on our HEP. With this in place we are acquiring a Twizy as part of the See More Programme

We will continue to develop Blencathra and encourage the positive environmental behavioural actions of both staff and visitors to promote carbon reductions and improved environmental sustainability.

GABRIELLE BACK

CREATIVE RECYCLING AND CLIMATE CHANGE SURVEYANCE

PROJECT AIM

- To address and raise awareness of several aspects of sustainability, including climate change and waste minimisation, through education, good practice and creativity
- To mobilise schoolchildren into researching and producing mini-projects on butterflies as indicators of climate change
- To engage schools, community groups and individuals to complete climate change surveys for analysis of local impact
- To encourage schoolchildren, community groups and individuals to address recycling and make a recycled butterfly depiction
- To involve schools in a recycled butterfly competition
- To create a giant recycled butterfly collage from submitted butterfly depictions for an exhibition to promote recycling and raise funds for Myton Hospice.

PROJECT ACHIEVEMENTS

- Several thousand butterflies as indicators of climate change surveys were completed
- Analysis revealed that some species were in decline in the area due to climate change
- More than 2,000 recycled butterfly depictions were made by schools, community groups and individuals
- A giant recycled butterfly collage was produced from around 2,500 butterfly submissions.

The collage is currently on display at Holy Trinity Church, an historic building in Coventry city centre. It is being viewed by hundreds of visitors and is raising awareness of recycling. The collage has raised more than £2,500 in sponsorship for Myton Hospice. The hospice logo is a butterfly.

WHAT DID THE PROJECT INVOLVE DOING?

Planet earth is a beautiful place which we inhabit, so we must constantly remind ourselves to focus on ways of reducing damaging environmental impact and encouraging sustainability. Raising awareness and engaging individuals to focus on aspects of sustainability can stop us harming our planet and changing the balance of nature. This project was inspired by the beautiful butterfly and uses it to raise awareness of recycling and climate change.

The project was developed and carried out in two parts:

- Education and environmental sustainability elements using butterflies as indicators of climate change
- Creative recycling focus, creating the giant recycled butterfly collage.

Environmental sustainability

- Schools were invited to research climate change and butterflies as indicators of climate change, and then to develop mini-projects
- Students, community groups and individuals were encouraged to complete climate change surveys using butterflies as indicators. More than 3,000 surveys were done and analysed
- Schools in Coventry and Warwickshire were encouraged to enter a recycled butterfly depiction competition. Butterflies could be made in any shape, size and medium using recyclable materials, waste papers, fabrics etc. Some 30 schools took part with over 2,000 submissions
- Creative workshops were set up to make recycled butterfly depictions with community groups during the Coventry Diversity festival. Participating schools received certificates and the winners were awarded cash prizes.

Creative and innovative recycling focus

- Some 2,500 recycled butterf[y depictions were submitted and the beautiful depictions ranged from 75cm to 35cm in size
- The butterfly depictions were compiled into a stunning giant recycled butterfly collage measuring 3.27 metres x 3 metres
- A civic reception was held to launch the collage display at Holy Trinity Church, one of Coventry's historic buildings in the town centre. The work will remain on display for the rest of the year.

WHY DID YOU DO IT?

Recycling and climate change are crucial environmental issues so it's important to raise awareness and engage people to address these aspects of sustainability through a creative and innovative challenge. I wanted to

involve schools in researching climate change and completing indicative surveys which were intended to inform on the effects of climate change on local wildlife.

Addressing waste minimisation through recycling is essential as many items are made from products that are slow[y running out. Some waste materials – particularly all forms of paper, card and fabrics etc – can be reused to create beautiful artwork.

The butterfly inspired this giant collage to raise awareness of recycling. Sustainability is key to our future survival and I

wanted to engage the public into focusing on eco-issues using creativity. The project extended participants' knowledge and community skills.

The project lent itself to some fundraising for a good cause, again enveloping an environmental aspect for providing funds for environmental improvements.

WHAT DID IT COST?

The project cost was in the region of £8,000, but there was a great deaf of voluntary contributions and work in kind. Research, development and organisation cost about £2,000. Production of stationery, leaflets, posters, postage and advertising came to £1,500 and £1,250 was spent on prizes and framed certificates. The launch and exhibition cost £1,500 and an additional £500 was spent on miscellaneous items.

The project was funded by Jaguar, National Savers, sponsorship donations, work in kind and volunteers. Holy Trinity provided a free exhibition venue.

WHO AND WHAT HAS BENEFITED?

This has been a fantastic project with a huge impact. It has involved thousands of participants during its different aspects. The climate change survey analysis revealed that climate change is indeed affecting wildlife species locally. There are fewer varieties of particular butterflies with some not being seen at all – clearly an indicator of impact.

Several thousand recycled butterflies were made and used in the creation of a giant collage within the centrepiece of a stunning exhibition, being viewed by thousands of tourists and raising awareness of recycling. The raising of £2,500 t in collage sponsorship for Myton Hospice environmental improvements is also a major achievement.

The community involvement and impact has been immense so the beneficiaries include all those who directly and indirectly participated, schools, community groups, individuals, all those involved in the creative activity and the audiences and viewers of the collage and participants in the exhibition.

The hospice for the terminally ill also benefits from the collage sponsorship and any environmental improvements to be made.The collage eventually is to be made into pictures to adorn the walls of the hospice. The long-term beneficiary is also the planet.

WHAT WAS INNOVATIVE?

This is a very innovative and creative way of engaging people into addressing environmental issues, raising awareness and in an unusual approach to getting public involvement and action for issues such as recycling. The fund raising aspect is also very novel, by using this themed approach it provokes thought and action.

IMPLEMENTING PROJECT METHOD

The project method can be repeated and duplicated, other environmental themes can be used. The recycled collage can he constructed using different reusable 3D materials. The exhibition can be launched every year at different venues. The collage can be used to raise money.

LESSONS LEARNT

This was a very extensive project and took longer than expected to complete. The surveys went very well and on time but the individual recycled butterfly depictions took a long time to come in from the schools etc so the competition closing date had to be extended.
The making of the collage was a huge task as it seemed to take forever to finish due to its huge size. There were problems with hanging it up on the stone walls of the church, so it had to he reinforced with special tape and

attached with metal pins. The project will definitely be repeated again but with a focus and theme.

GIRAFFE INNOVATION

VIRGIN ATLANTIC ECONOMY MEAL SERVICE

PROJECT AIM

The aim of the project was to reduce the environmental impact and costs associated with the economy class inflight meal service within the context of Virgin's customer experience and brand efficacy.

Working with the Virgin Atlantic design team, MAP design and procurement teams, the project involved a detailed analysis of every aspect of the design including functional/service requirements, material type, material quantity, material density, component lifespan and recyclability.

The project aimed to make changes that would reduce the environmental impact and associated costs of the meal service, while leaving a lasting legacy of the value of integrating lifecycle thinking into the early stages of design and procurement decision-making.

This would all need to be achieved while maintaining the core design principle of cost, aesthetics, function and brand and providing an improved user experience.

The project sought to challenge perceptions that the only major design factor worth considering with regard to the environment was 'weight' related to fuel burn. Although weight is an important element in considering environmental impact, material type, embodied energy, manufacturing process, disposal versus reusable (rotable), sector life (routes flown) and end of life recycling also had a major impact on overall greenhouse house gas (GHG) emissions.

PROJECT ACHIEVEMENTS

The project achieved the following benefits:

- £ 8.63 million annual saving
- 4% overall CO2 reduction on inflight service
- Weight saving – 129kg per aircraft with a fuel saving of 762 tonnes (2,400 tonnes of CO_2) a year

Meal appeal: serving up changes

- 10% reduction in waste
- 52% increase in recycling
- 25% reduction in storage required
- 9% improvement in customer satisfaction.

This project highlights the benefits of factoring environmental analysis and eco-design principles from the outset and that aesthetically pleasing, functional, economic designs are not mutually exclusive to an enhanced environmental performance.

The project introduced recycled material content to specific inflight meal service items. The tray was changed from acrylic to 100% recycled ABS (rABS), and the cups were changed from Virgin PET to 100% recycled PET (rPET).

These changes were complemented by the all-important reduction in weight and with an enhanced brand appeal. A white paper tray liner was replaced by a soft over-moulding which reduced weight further and eradicated the paper waste.

Where a set service life can be achieved, the modelling showed that heavier more durable reusable items reduced the environmental impact over the components whole life.

WHAT DID THE PROJECT INVOLVE?

Giraffe engaged the VAA in-house design team at the early stages of the design process to highlight the opportunities of eco-design and life-cycle analysis (LCA) in the product development process.

The initial stage involved lifecycle modelling of different materials types over different sector lives (routes) to highlight that material choice and end of life options can also have a major impact on environmental impact and costs.

The modelling at this early stage acted as a mnemonic to guide design decision making throughout the development process. Giraffe's key involvement included:

- Lifecycle modelling which influenced the design and decision making process, including material selection, density and manufacturing process
- Recycled plastics (rplastic) options and recyclability
- Sector life – indicative routes
- Disposal versus reusable items – lifecycle and break even impact
- End of life – recycling.

Giraffe also engaged key suppliers in ensuring that specific design changes (i.e. rPlastics) could be made within cost. Initially one key supplier stated that recycled plastics "couldn't be done". Giraffe used their technical knowledge and case studies to show it was achievable at no extra cost.

WHY DID YOU DO IT?

The project started because the meal service had not had an overhaul for 15 years and was being delivered to around 4.5 million customers annually. It was part of Virgin's £100 million investment program for in-flight services.

The project highlighted the importance of factoring in environmental thinking from the outset. VAA wanted to use the analysis to reinforce the business case for the new in-flight service and gain internal buy-in to the subsequent investment programme.

Having a demonstrable environmental and cost benefit along with a high level of customer experience were seen as part of the same package. The project also support the VAA Change is in the Air sustainability strategy and reporting.

WHAT DID IT COST?

The combination of service design and logistical improvements was a compelling case for the Virgin Business Unit to become a key element of Virgin's £100 million investment program for in-flight services over the last three years. The project budget for the environmental analysis and advisory from Giraffe was £20,000.

WHO AND WHAT HAS BENEFITED?

The success criteria for the project included improving cabin crew's ability to better serve customers and improve passenger experience of the meal (measured by surveys). The design achieved a 9% increase in customer satisfaction and will ultimately benefit approximately 4.5 million passengers a year.

WHAT WAS INNOVATIVE ABOUT THE PROJECT?

Virgin's in-house team developed an improved service idea where the dessert plus tea or coffee was served as a separate course after the main meal, which is more akin to eating in a restaurant.

This improved customer experience but also created extra space on the meal tray. The subsequent tray design was 30% smaller which meant the meals could be arranged in the meal trolley in rows of four instead of three, saving 25% in storage requirements overall.

Several trolleys could be removed from the aircraft, thereby saving weight. The trays include a lip so they hook together. When the cabin crew pulls one meal out, the next tray slides forward so they don't have to reach into the back of the cart.

A soft plastic non-slip inner was co-moulded into the tray to provide colour and stop product movement. This meant the traditional paper liner could be removed, saving weight and reducing waste. This collaboration wasn't just about reducing emissions, but doing so within the commercial context of cost and the Virgin Atlantic brand.

Where a set service life can be achieved, the modelling showed that heavier more durable reusable items reduced the environmental impact over the components whole life. For example, the long china dish achieved a 10% reduction in carbon footprint over the sector life (number of uses), despite a 16% increase in weight.

This project is innovative as it changed multiple aspects of the design to materials that had not previously been considered and disrupted assumptions about material use and reuse versus disposable options.

This innovative project delivered a design solution that both reduced environmental impact, saved the company money and offered an improved user experience.

WHAT ARE THE LONGER TERM BENEFITS?

The new design was widely praised in the press for delivering a better service, more attractive design and reduced environmental footprint.

In terms of longer term benefits, Giraffe Innovation also delivered cross departmental workshops to train teams in sustainable design and procurement practices, and how to work positively with suppliers to improve standards.

Virgin increased the number of suppliers signed up to their Sustainable Procurement Policy from 26% to 50%. This training and the successful reduction in the environmental impact of the design have left a lasting legacy with the Virgin Atlantic design and procurement teams of what can be done to reduce cost and environmental impact.

GOVARDHAN ECO-VILLAGE, INDIA

GEV SYMBIOTIC DEVELOPMENT MODEL

PROJECT AIM

Modern developmental systems have been formulated on the premise that nature has abundant resources and have hence adopted a linear model of "resource to-trash" systems. They take resources from nature, process them into a useful form and at the end of life-cycle dump it back into nature in the form of waste.

With more nations of the world adopting this model, there is an increasing strain on the planet's ecological resilience resulting in effects like global warming, climate change, loss of biodiversity and food shortage etc.

The aim of Govardhan Eco-Village (GEV) is to solve this environmental crisis by becoming a global template of sustainable eco-friendly community for the whole world. The innovative approach is that of having symbiotically dependent systems based on the following principles:

- Nature has a very unique way of waste management. The systems comprising the developmental model in nature are in such a way that the wastes from one system can be used as raw materials for another system
- By aligning our developmental patterns of modern amenities with this inbuilt symbiosis system, we can ensure a sustainable future.

PROJECT ACHIEVEMENTS

- 500 tribal families from six villages in the Wada block of Thane District benefited from improved agriculture through organic farming, water resource development, self-help groups, skill enhancement and help for landless families
- A ten-million litre pond was excavated to harvest rain water and reduce river water dependency
- A soil bio-technology (SBT) System was established to recycle waste water with the capacity of 30 KLD
- At GEV we follow sustainable organic farming that is beneficial for the existing biodiversity as well
- GEV has successfully demonstrated a 30 cubic metre model biogas plant, which runs on cow dung and other food wastes.

GEV also houses a 30kVA solar power plant and many standalone solar powered street lights which minimize its dependence on fossil fuel based power sources.
GEV also houses various animal driven prime movers (ADPMs) which utilize animal power such as bull driven water pump, bull driven grass cutting machine and various bullock carts used for transportation.

We have employed mud based technologies like Compressed Stabilized Earth Block with less than 1% embodied energy compared to conventional buildings. Today GEV is a proud place to showcase a Symbiotic Development Model.

WHAT DID THE PROJECT INVOLVE DOING AND WHY DID YOU DO IT?

GEV is an integration of various symbiotically dependent sustainable initiatives. These include:

- Water conservation and the harvesting of ten million litres of rain water
- Sustainable organic farming spread over an area of more than 14 acres
- Rural specific alternative energy sources consisting of a 30 cubic metre biogas plant, a 30 kVA solar power plant and stand alone solar powered street lights
- Various animal driven prime movers (ADPMs), which utilize animal power such as a bull-driven water pump, bull-driven grass cutting machine and various bullock carts used for transportation
- A waste water treatment technique called the soil bio-technology (SBT) plant
- Green buildings utilising the concept of mud -based technologies such as compressed stabilised earth blocks (CSEBs), Cob and Rammed Earth built structures that have less than 1% embodied energy compared to conventional buildings.

All these initiatives work in harmony with each other, with the waste product from one system being used up as the raw material for another system, thus creating a sustainable cycle. Our purpose is to develop farm communities as they present the only sustainable way of leading a life which is physically, mentally, socially, economically, spiritually and environmentally friendly. GEV showcases the profound principle of 'simple living, high thinking' and its relevance in today's world

Our organisational model is based on corporate funding and funding from individuals. Our fundraising is planned through CSR activities of Businesses. Corporate Visitors who visit our eco village often are source of our income. Well-wishers also make contributions in the form of donations.

WHO AND WHAT HAS BENEFITED?

A ten million litre pond was excavated to harvest rain water and reduce river water dependency. A soil bio-technology (SBT) System was established to recycle waste water with the capacity of 30 KLD. GEV has successfully demonstrated a 30 cubic metre model biogas plant, which runs on cow dung and other food wastes.

With our Integrated water conservation programme and waste water Recycling we have:

- Reduced our dependency on pumping of river water from long distance after implementing the results of this report
- The landscaping water requirements are met through increased underground water levels
- 100 % of treated water is being reused in horticulture. The revenue from horticulture comes to Rs 1.5 Lakh per annum
- The by-products from waste water treatment system based on SBT are biomass and bio-fertiliser
- A maximum of two metric tons a year of bio-fertilizer can be generated from the system as a by-product
- There is no toxic waste and there is no foul smell from our systems.

WHAT ARE THE LONGER TERM BENEFITS?

Govardhan Eco-Village is an internationally famed award winning farm community that has set the best practices in diverse fields of water conservation, waste management, alternative energy, green buildings, organic farming and rural empowerment.

The many long term benefits include:

- More communities have benefited
- Visitors coming to GEV get inspired by our eco-friendly activities and are motivated to carry them out. Annually more than 15,000 students visit the GEV and learn about the various eco-initiatives
- More strategically important partnerships with government and technical institutions
- Increase in biodiversity
- Reduction in overall carbon footprint and an increase in carbon sequestration through plantation
- Increase in profit.

WHAT IS INNOVATIVE ABOUT THE PROJECT?

The symbiotic development model in Govardhan Eco-Village is an innovative model whereby it is blending the wisdom from traditional practices and the practicality of modern science.

This model, unlike the modern development systems, is cyclic in nature with the waste coming out from one system goes as raw material for another system. thus saving the energy for processing of waste.

An opportunity was provided for employing an eco-friendly concept called Soil Bio Technology (SBT), developed and Patented by IIT Bombay. The plant was developed by Mr. Biplab Patnaik (IIT-B) of Lifelink technologies.

GEV was the first successful rural project to implement this technology. A hydrological survey was conducted by the Advanced Centre for Water Resources Development and Management (ACWADAM) to find out the best options for water harvesting, recharge and discharge areas at GEV; which suggested the excavation of the ten million litre pond.

With the pond funded by Government the greenery around it has increased very much. Biodiversity of the region has been observed to be increasing. Working in collaboration with Bharatiya Agro Industries Foundation (BIAF) we have been conducting agriculture training camps for farmers in many villages, which can serve as good field visit sites for the participating youths.

WHAT HAVE YOU LEARNT?

We have learnt that like all community campaigns, a major risk to our activities is in ineffective publicity, lack of participation and not sustaining enthusiasm of the participants all throughout the project life. In order to cater for this, we have already tailormade the model to suit the needs of the communities and understood the results. We plan for a horizontal scaling up such as expansion and replication. We have established strategic alliances with key stakeholders that can enable us to scale up the system at various places.

Over the past few years, Govardhan Eco-Village has been making steady attempts to generate positive change waves on the social and environmental front through its multifarious programs and projects.

This has been confirmed by the recent recognition and awards GEV has received from various quarters. In 2015 the India Chapter of the International Advertising Association (IAA) awarded GEV the IAA Olive Crown Award for Creative Excellence in Communicating Sustainability.

GEV also received the Golden Globe Tigers Award for Best Green Water Stewardship in the same year.

GREEN MOTION

LOWERING CO_2 EMISSIONS

PROJECT AIM

Green Motion car rental is concerned about lowering CO2 emissions associated with vehicle travel and also with protecting the environment through the employment of recyclable materials – stationery, eco-friendly cleaning products and recycling rubbish.

The lowering of our use of CO2 emissions and our use of sustainable material/products is not simply a project for Green Motion. Rather, it is an intrinsic part of our brand and values. Currently, Green Motion is the only completely eco-friendly vehicle hire company, offering its services to both business and leisure clients.

Green Motion, the pioneers and leaders of the vision of low carbon vehicles for daily rental, are going from strength to strength since the company was formed in 2007. With carbon reduction at the heart of the company, Green Motion has remained focused on continued evolution, ensuring it maintains its position as number one in the provision of low CO2 vehicle rental.

PROJECT ACHIEVEMENTS

For us, reducing our impact on the environment isn't just good for business, it is our business. All of our vehicles abide by a strict CO_2 emission guide as Green Motion boasts of maintaining the greenest and cleanest vehicle rental fleet in the world. Its collective vehicle emissions stand at sub 105g/km which is more than 25g/km lower than its closest competitors.

We also offset our carbon usage by our Green Heart Fee initiative, which allows customers to donate to the Costa Rican rain forest. So far we have donated around £30,000 to the Costa Rican government which has helped to plant around 10,000 trees.

Green Motion staff also work with customers and employees on their travel strategies, educating them on eco-friendly driving tips to ensure that our legacy with our customers does not end at the conclusion of their rental.

WHAT DID THE PROJECT INVOLVE DOING AND WHY DID YOU DO IT?

Green Motion was formed in the UK in 2007 by entrepreneur, Richard Lowden. It has since grown into an international company operating in 19 countries (and growing). The company is based on an eco-friendly ethos and we are pioneers and leaders of the vision of low carbon vehicles for daily rental.

As mentioned, the lowering of our use of CO_2 emissions is not simply a project for Green Motion; rather, it is an intrinsic part of our brand. Being the only rental company that is genuinely committed to its environmental ethos, Green Motion has been designed to constantly evolve, staying ahead of its competitors by offering a completely unique product within a highly congested and competitive arena, ensuring that we instantly stand out from the crowd.

For us, reducing our impact on the environment isn't just good for business, it is our business.

The Green Motion model is unique as it does not charge a premium to go green and genuinely delivers on its environmental policy, educating customers that they do not have to compromise on cost or comfort to go green.

Our model has been crafted to attract not only the ecologically conscious individual but also customers who may choose Green Motion's services for its competitive vehicle hire rates and excellent customer service.

Branching out: Green Motion offers the new BMW i3 Electric for hire

The company's innovative objective is simple – to offer customers (both leisure and business) the opportunity to rent a vehicle from a company that genuinely cares for the environment. Although our business endeavours to lower its usage of CO_2 emissions associated with road travel, Green Motion also continues to concentrate on its own sustainability by remaining dedicated to further reductions in energy use, minimising water consumption, effective recycling and the employment of sustainable and eco-logical materials throughout our business platform.

CO_2 emissions affect not only our climate but also our health. How we treat our environment in the short-term has detrimental consequences in the long term. We believe that all businesses can put positive infrastructures in place to care for the environment.

As a car hire company, we recognised that we had a responsibility to care for our environment in a market that is known for not being entirely environmentally friendly.

The key stakeholders of our company are the people who invest in our company – our franchisees, our customers and our staff. We challenge them to make behavioural changes at work and at home.

In the short term, we are helping to reduce the impact that our business activities has in providing CO_2 emissions. In the long-term, we are hopeful that as our brand continues to strengthen and grow, our presence will help to inspire other companies and individuals to make behavioural changes and start doing their own bit for the environment – viewing ourselves as a collective team rather than competition.

The company runs on a franchise system and each franchised location must adhere to strict guidelines for the rental vehicles they offer to customers. Green Motion is proud to have the greenest vehicle rental fleet in the world: our collective vehicle emissions stand at sub 105g/km which is more than 25g/km lower than our nearest competitors.

Additionally, our franchisees must also be conscientious regarding the materials that they use – this relates to minimising water consumption, effective recycling and the use of sustainable and eco-logical materials throughout our business platform.

Our franchised locations are all true ambassadors for the green vehicle movement and they regularly inspire us to continue to make changes to the way we view our business.

The master country franchisee for Green Motion Costa Rica launched the Green Heart Fee initiative in 2013. The Green Heart fee is a fee which is charged to all customers, but they have the option to opt out. The fee is then donated directly to the Costa Rican government who deploy the funds for environmental causes, including the re-forestation and protection with the rainforest of Costa Rica.

To date, the combined efforts of Green Motion and its customers have achieved a position of carbon neutrality by donating roughly £30,000 which has helped to plant circa 10,000 trees. Green Motion is excited when our franchises create innovative ways to accelerate our environmental causes. As such we have proudly sponsored the Green Heart Fee since its conception and have started rolling the initiative out to all of our other 19 countries.

Additionally, our Green Motion Netherlands franchise has also demonstrated their commitment to caring for the environment by investing in a machine which enables them to clean their vehicles with steam. This allows them to save up to 90% on their regular usage of water, saving more than 5 million litres of water per year (and reducing their water bill – being eco friendly doesn't always mean you have to spend more).

The image that our rental branches advertise to customers must also embrace the sustainable ethos of our brand. Customers are encouraged to share our brand values during their rental as our staff provide eco-friendly driving tips and involve customers in the Green Heart Fee. We also try to stay connected with customers after their rental through our social media sites.

Our brand is proud to be an award winning car hire company – we have won Green Fleet's 'Rental Car Company of the Year' five years in a row (2009 to 2014), Energy Saving Trust's 'Fleet Hero Award' in 2011, the 'Environmental Rental Company of the Year' for the Institute of Transport Management* in 2011 and most recently, Green Motion was awarded a National Silver in the Medium Business category for the Scoot Headline Awards for 2015.

The proof that our company is impacting our stakeholders (franchisees, customers and staff) remains evident by our continued success. Green Motion is absolutely committed to the ongoing development of its unique brand and positioning with further expansion underway in North America (including the United States and Canada), along with Australia, China and New Zealand.

GROSVENOR SHOPPING CENTRE, NORTHAMPTON

SUSTAINABLE CENTRE GARDEN

PROJECT AIM

To deliver an sustainable garden area in the middle of a shopping centre, which also sits in the middle of a busy market town. The area had been a garden in the past, but this had been overlooked and neglected for many years and had become overgrown and almost inaccessible.

The project was the idea of Russell Hall, who at the time had recently joined the centre team as the most senior employee at the centre. The project had three key purposes:

- To enhance team building amongst staff
- Deliver an environment which would allow local habitat to flourish
- Deliver an area in which staff could be proud of their achievement.

PROJECT ACHIEVEMENTS

The vision was to never to simply create a quiet area. We wanted to expand upon this by developing an environment, which would allow a platform for wildlife, flora and fauna to flourish, while also enabling staff to participate in an exercise to develop team building and take a sense of ownership.

We were committed to using recycled materials to deliver the environment, using water from surrounding roof areas (grey water), purchasing plants from local nurseries (or in some cases growing direct from seed), growing vegetables and fruit for staff to take home and share with their families and friends. We were also demonstrating how easy it is to grow your own produce, whatever the environment.

WHAT DID THE PROJECT INVOLVE?

The Grosvenor Shopping Centre in Northampton was originally constructed in 1975. The centre has two separate blocks of office accommodation, which sit on the service deck above the centre and are both currently unoccupied. Between one of the above office blocks and the delivery/service deck, albeit totally enclosed, is what was once an enclosed garden area. After many years of neglect this had become overrun and neglected.

Blooming: the Grosvenor garden

The plan was to turn this area into an external environment which offered the potential for staff to find a quiet area to get away from the usual hustle and bustle associated with a shopping centre environment.

WHAT DID IT COST AND WHERE DID THE MONEY COME FROM?

The exact costs are minimal, insofar as these were re-directed from monies previously spent with the local council for supplying floral displays around the centre.

The centre has a small CSR budget from which some of the more costly items were funded. These included:

- Beehive and bees
- A wormery that produces concentrated food feed and compost from staff food waste
- Seating
- Bird boxes and butterfly rests
- Composting bin
- Water harvesting – water was taken from surrounding roof areas.

The estimated cost was around £750, which was mitigated by members of staff voluntarily giving 100 hours of their time. The beekeeping is undertaken by two members of staff. Unlike other shopping centres which have hives and outsource the beekeeping, the Grosvenor Centre undertakes the role in-house.

WHAT DID THE PROJECT ACHIEVE IN TERMS OF SUSTAINABLE DEVELOPMENT, ECONOMY, ENVIRONMENT AND/ OR EQUALITY?

This is only a bargain basement exercise, but the buzz it has created is unquantifiable. Quite simply it is there for all to enjoy.

Sustainable development

It proved an overrun garden in a town centre surrounded by large concrete buildings can thrive as an eco-garden and be enjoyed by those associated with the working area and its guests.

Through the commitment of staff, the area will continue to thrive and hopefully be made available for school groups and others to visit as part of additional changes proposed for the centre.

Economy

It is not a project to enhance the local economy to any great level, although our bees and associated beekeeping equipment were purchased from a local beekeeper. Our garden seating, bird boxes and butterfly rests were purchased from a local charity, St Andrew's, who specialises in training people with severe disabilities. Replacement plants were purchased from local nurseries.

In the bag: greenery in the centre of town

Environment

The project has greatly enhanced a rundown and neglected area and turned it into an environment which has promoted a thriving honey bee colony, provided housing for small birds, and is accessible for all.

Equality

Available for all, to be enjoyed by all.

WHO AND WHAT BENEFITED?

All in-house staff, and a number of retail staff employed in the centre. School groups will hopefully benefit going forward, and not to forget the local wildlife community.

WHAT ARE THE LONGER TERM BENEFITS?

Creating an environment in which local wildlife can thrive. prosper and multiply

WAS THERE ANYTHING INNOVATIVE ABOUT THE PROJECT?

A simple project that was achievable using a limited budget, but one which relied on the passion and drive of staff. This area would have remained neglected without the vision to deliver into what is now a very special area.

CAN OTHER ORGANISATIONS BENEFIT FROM IMPLEMENTING YOUR METHODS?

Anybody who has a limited number of staff and a green space/area can, regardless of location, undertake a similar project to turn something which may have been neglected and forgotten about into a truly amazing environment that can be shared by all.

WHAT DID YOU LEARN?

Staff now have the bug.We've recently completed a green wall, solely using discarded pallets and plants from local nurseries. Our next project which we are just finalising is to introduce a 'green roof', which will be the first in the town centre and be visible from the centre's scenic lifts

We are confident this is very much the tip of the iceberg, staff are now motivated to achieve more for the community and the environment. For a nominal cost we have brought together a real sense of bonding and pride amongst staff and delivered an environment in which all associated can be proud of their achievements.

NEWLY REFURBISHED
WITH OVER
50 HIGH STREET
RETAILERS!
TOPSHOP
極度乾燥(しなさい)
Superdry.
Boots
NEW LOOK
PRIMARK®
next
GROSVENOR CENTRE
G
GROSVENOR CENTRE
NORTHAMPTON
GROSVENORSHOPPINGCENTRE.CO.UK

HANG TUAH JAYA MUNICIPAL COUNCIL, MALAYSIA

MANAGING CARBON EMISSIONS

PROJECT AIM

Hang Tuah Jaya Municipal Council or Majlis Perbandaran Hang Tuah Jaya (MPHTJ) has carried out several projects to manage carbon emission within the organisation's administrative area which includes MPHTJ's Low Carbon City Framework (LCCF) and Carbon Footprint.

The LCCF aims to produce a carbon emission baseline, collect existing carbon emission data, propose carbon emission reduction strategies and recommendations, as well as annually set a carbon emission reduction target.

The Carbon Footprint program aims to produce carbon emission baseline, collect/investigate existing carbon emission within the organization, carry out SWOT analysis (strengths, weaknesses, opportunities, threats) in order to propose carbon management strategies and recommendations and set an annual carbon offset target.

PROJECT ACHIEVEMENTS

Carbon emission baselines were successfully collected from year 2012 until today. In fact the carbon emission baseline data collection is still in progress since it is an ongoing process. In 2014, carbon emission baselines show positive results. It was found that there was a decrease in carbon emission to 434.26 tCO_2 as compared to 2013. This indicates a 4.30% reduction in carbon emission.

This was due to the awareness programs and green practices within the organisation which encourages energy conservation from the usage of electricity, water and other sources. The Ministry of Energy, Green Technology and Water Malaysia (KeTTHA) has presented MPHTJ with a diamond award that indicates Best Practice 1 (1% 9% carbon reduction level).

Carbon footprint

Based on the data collected, at least 810.68 tCO_2 was produced within the organization in 2013. The main sectors that contribute to this number were from the use electricity, solid waste, papers, fuel and water. To offset the carbon produced in 2013, MPHTJ has planted 811 trees, which has led to successfully compensating the carbon produced in 2013.This practice continues to be carried out annually according to the findings.

WHAT DID THE PROJECT INVOLVE DOING?

In line with the aspiration of the state's Chief Minister in making the State of Melaka a Green Technology City, the Hang Tuah Jaya Municipal Council or Majlis Perbandaran Hang Tuah Jaya (MPHTJ) has carried out several pilot studies to manage carbon emission within the organisation's administrative area which includes MPHTJ's Low Carbon City Framework (LCCF) and Carbon Footprint.

MPHTJ was among the pioneer local authorities to sign a Memorandum of Understanding (MoU) with the Ministry of Energy, Green Technology and Water Malaysia (KeTTHA) and the Malaysian Green Technology Corporation (MGTC). This was to enable the implementation of LCCF in office buildings located within the administrative area of MPHTJ.

MPHTJ's LCCF initiative was implemented on 2012. A RM50,000 fund has been given by KeTTHA to implement this initiative in MPHTJ. Among the organisation's internal departments involved in the execution of this study are the Department of Town Planning, Department of Landscape, Department of Development Control, Department of Engineering and the Department of Buildings. This study involves 14 administrative buildings within an area of 1,956.57 acres.

WHO AND WHAT HAS BENEFITED?

This initiative is one of the initial steps taken by MPHTJ to create awareness among its members on conserving energy and resources. This effort indirectly proves that MPHTJ can make a difference towards creating a better environment and set an example to be emulated by other stakeholders (residents, developers, planners, etc).

In addition, various awareness programs and green practices have been carried out in the 14 administrative buildings.

As a result, the rate of carbon emission in 2014 decreased by 434.26 tCO_2, which is equivalent to 4.30% reduction, compared to the readings recorded in 2013. This can be compared to saving RM739,379.00 worth of electricity bills. By practicing conserving energy, this benefits building users particularly in terms of reducing monthly financial cost.

In the long run, this program is able to improve the saving ratio through green practices and awareness programs. MPHTJ plans to continue expanding the scope of research and study of LCCF to include all buildings within the administrative area of MPHTJ.

MPHTJ also carries out the Carbon Footprint study within its organisation. MPHTJ is the first local authority in Malaysia to take up this innovative step. Being aware of the need to control and manage carbon emission from daily activities within the organization has led to initiating this study.

It involves collecting primary and secondary data such as meter readings, electricity and solid waste generation data, inventory and audit of electrical equipment, utility bills, vehicle records, asset records, and reports of the study (data collection is a continuous process).

These data are useful in planning and designing effective strategies for implementation. In terms of cost implications, the study did not involve any particular costs because it is carried out within the organization and it is expected to be expanded into the next phase. This study focuses on managing the amount of carbon emitted within the organization through employing the carbon offset approach. In 2013, it was found that 810.68 tCO_2 has been produced from five key sectors, namely energy, solid waste, paper, fuel and water.

Hence, largescale tree planting program was carried out to offset the amount of carbon emitted. A total of 811 trees have been planted, specifically intended to offset the carbon emission, assuming that a broad leaf tree will absorb in the region of one tonne of carbon dioxide during its full lifetime. The organisation benefits from this program because this creates awareness on resource conservation and green practices in the everyday work of the organization.I n the long run, this program aims at incorporating the involvement of other organizations and communities under the administration of MPHTJ.

Through this program, greater awareness on the importance of carbon management can be established and MPHTJ could act as a role model to a community of 161.290 people in its administrative area.

Throughout MPHTJ's four years of involvement and commitment towards nurturing sustainability and green practices, the organisation has received various awards. Among the honours and awards received include Green Neighbourhoods Initiative Award 2011 (walkway), Office Green Practices Award 2011, Green Neighbourhoods Initiative Award 2012 (cycle paths), Sustainable City Award (by MURNInet) 2012 and first runner-up in the Green Practices Program Melaka State Level 2012.

In conclusion, the implementation of both the highlighted programs have allowed MPHTJ to create awareness on sustainability and energy conservation to members within the organisation as well as the community in general.

This brings the organisation one step closer in realising the Melaka State Government's aspiration of establishing Melaka as a Green Technology City.

HM REVENUE AND CUSTOMS (HMRC)

HMRC PAPER REDUCTION PROJECT

PROJECT AIM

The aim of this project was to reduce substantially HM Revenue and Customs' (HMRC) environmental impact and costs by cutting paper usage during the five-year period 2010-15. The aim supported the Greening Government commitment to reduce paper usage. It was aligned with HMRC's business objectives to improve the customer experience, embrace digital agendas and cut costs.

The project included both the volume of paper used internally by our 56,000 full-time equivalent employees, for example in printing and photocopying, and the paper used to deliver information and services to our 45 million individual and 4.9 million business customers.

Our objectives were to:

- Change employee behaviours by encouraging them to print less and print double-sided
- Meet customer and stakeholder expectations on how they receive information from HMRC
- Use an array of communication channels to develop and encourage greater use of electronic customer interaction
- Use a system of 'closed-loop' recycling so that waste printer and copier paper is recycled into new paper for our machines
- Save money on paper purchases and cut waste.

Success was measured through:

- Reductions in paper used in both internal and external communications
- Customer take-up of electronic communications
- Volume of paper put through the 'closed-loop' system cost savings.

project achievements

- 170 million fewer sheets of paper used by employees when printing, during 2010-15
- Responded to the 97% of stakeholders who, in our annual stakeholder survey, said they would like email contact with us, and used customer communication channels such as: poster adverts; radio; online; email; text; twitter; digital escalator panels and bus shelter panels, all of which resulted in:
- Half of the 11 million people who file assessment returns opting to receive correspondence from HMRC digitally, as the first step in cutting 31 million pieces of Self -Assessment paper each year
- One million small businesses signing up to an outbound email service, massively reducing paper-based correspondence
- VAT returns printed each month reducing by 85%, from 200,000 to just over 30,000, as part of a campaign to encourage electronic filing
- HMRC was the first government department to introduce a 'closed-loop' system for paper recycling, with more than 3,500 tonnes of paper put through the system annually
- Procured 170 million fewer sheets of paper for internal use, at a saving of more than £864,000
- Contributed to a reduction in waste generated of 10,000 tonnes.

WHAT DID THE PROJECT INVOLVE AND WHY DID YOU DO IT?

Campaigns and coordinated communications to change behaviours, cut paper usage in internal operations and external service delivery, and to persuade other parts of central government to recycle waste paper through a 'closed-loop' system.

- Paper is one of HMRC's biggest outputs and a major environmental overhead
- Lord Carter of Coles' review of HMRC's online services recommended universal delivery of tax returns by 2012.

In addition, we wanted to:

- Recognise environmental and costs savings afforded by our unique position as one of the largest government departments, with a customer base extending to almost every adult in the UK
- Address reputational issues associated with paper usage
- Encourage our workforce to adopt more environmentally, sustainable and cost-effective ways of working
- Meet our customers' expectations by offering them first-class online services
- Reduce paper usage by 10% for 2011-12, in response to the Greening Government target
- Allow other government departments to benefit from our pioneering 'closed-loop' paper recycling process.

WHAT DID IT COST?

The costs were negligible:

- Online service initiative and tax compliance behaviour costs have been excluded
- Closed-loop manufacturing costs are incorporated within recycled paper costs.

WHO AND WHAT HAS BENEFITED?

- HMRC, from £864,000 savings in paper procurement costs and waste; these savings benefit the taxpayer, by making better use of public money
- Our customers, from the shift to online services, which improve accessibility and provide services at a time to suit them, and reduce costs for businesses by simplifying and speeding up contact with us
- Our employees, by being able to monitor their printing and use online tools that reduce the need for printing
- Other government departments, from having access to a 'closed-loop' system for paper recycling.

WHAT WAS INNOVATIVE APPROACH ABOUT THE PROJECT?

Within HMRC:

- Our Executive Committee-level Environment Champion provided visible senior leadership
- We set annual paper reduction targets across HMRC and for each individual business area. Our Executive-level Performance Committee monitored progress monthly and we developed interactive 'Hub slides' and data analysis for business leads to pinpoint areas of risk, for example single-sided, colour or A3 printing
- We delegated budgets for paper procurement to business areas, so that costs of paper use are felt more keenly at a local level
- We introduced a Managed Print Service, which defaults to double-sided printing, and dual-monitor computer screens which enable staff to work on multiple documents, eliminating the need to print
- We harnessed the influence of 500 'Green Volunteers' from our 170 offices across the UK, to spread the 'think before you print' and 'share papers at meetings' messages. Other communications included a staff poll to find out why people print, combined with 'print cold-turkey weeks' on HMRC's intranet, where teams pledged to go print-free
- We developed an Environmental Waste Calculator to enable employees to calculate the cost of using paper and printing and to identify the financial and carbon savings from reducing paper usage
- We produced a quarterly online magazine – The Green Room – to encourage staff to use fewer resources, including paper, and used environmental 'events' such as Climate Week, to provide tips.

Externally, in service delivery:

- HMRC's 2014 media campaign, 'I found inner peace when I did my Tax Return online ...' incorporated UK-wide print, radio, online and digital outdoor advertising, encouraged taxpayers to interact with us electronically.
- We reviewed 1,600 customer forms, scrapping 550 of these entirely and converting 190 to online i-Forms
- 650,000 customer interactions through webinars, instructional videos and apps reduced the need for paper-based communications
- We redesigned letters and forms to fit on fewer sheets of paper
- We moved the HMRC website to GOV.UK to provide a fully-integrated online service. We receive more than 50 million hits per month, many of which are saving on letters, printed leaflets and paper correspondence.

Externally, across government:

Case studies and impactful presentations to other departments led to the 'closed-loop' system being mandated across central government.

CAN OTHER ORGANISATIONS BENEFIT?

We've shared our experience with other government departments through case studies in reports on performance against government targets, encouraging others to follow our lead.

As the first department to introduce a 'closed-loop' paper recycling scheme, we led the way in encouraging 17 other departments to sign up, ensuring that waste paper is transformed into desktop-quality recycled paper, for reuse. Environmental benefits are 60% less energy and 50% less water than 'virgin' white paper.

From individuals to small businesses and multi-nationals, HMRC's customers have benefited from digital information: simplifying and speeding up the way they deal with us, and making it easier for them to meet their tax obligations.

WHAT HAVE YOU LEARNT?

1. The importance of senior-level leadership in the form of an Environment Champion, to give environmental issues visibility at board level.

2. The importance of regular monitoring and reporting against targets, to identify where further action is needed and to celebrate progress.

3. The positive impact of providing simple steps that everyone can take to reduce paper, for example:

- Using the 'print preview' function to print only the required parts of a document
- Making sure email addresses and signatures run horizontally rather than line-by-line down the page
- Storing information on shared computer drives, to reduce the need for printing.

4. The scale of the environmental impact, for example the reduction of 170 million sheets of paper for internal usage equates to saving approximately 20,000 trees.

ARE YOU PLANNING ANY FURTHER DEVELOPMENTS ?

- We're working to transfer all our tools and tax calculators to GOV.UK and to expand our Twitter customer service. We have 40 digital projects in the pipeline to save paper and to provide improved customer service and employee experience.
- Introduction of 'Google Share' to allow multiple editing of documents and savings in version printing.
- A 2015-16, 5% paper reduction target for each business area and an HMRC 3% waste reduction target. Performance will be monitored by HMRC's Executive-level Performance Committee.

HIGH CHELMER

PV SYSTEM DESIGNED TO PROVIDE SOLAR POWER

PROJECT AIM

High Chelmer carried out a building project in July 2011 that increased the main Primark anchor store from 37,000 sq ft to 61,000 sq ft. The building was completed and handed over to Primark, which opened the store in September 2012.

High Chelmer then arranged for JoJu (solar panel specialists) to install 17.1kWh PV solar panel units, fixing accessories to the roof area and associated logistics in June 2013 on our new Primark store.

High Chelmer strives to ensure that wherever possible we implement effective and innovative environmental strategies and measures to reduce consumption and enhance energy efficiency across the common environs of the centre, enhancing our CSR status in the wider community.

Our Client, LaSalle Investment Management on behalf of Coal Pension Properties Limited, and our management company Capita were keen to achieve ISO 14001 Environmental Management certification and a high BREEAM rating.

The aims of this project were to:

- To generate electricity for the national grid for others to use whilst making use of unused roof space
- To assist our client and management companies achieve ISO 14001 Environmental Management Certification and a high BREEAM rating for the project
- With other initiatives, reduce the centre's energy costs and carbon footprint to benefit tenants.

PROJECT ACHIEVEMENTS:

Inspired by a smaller solar panel water heating project installed on the roof of our six-storey office complex, High Chelmer called Threadneedle House in early 2012. A recommendation was made to consider a larger scheme for the Primark building project ultimately achieving the following to date: -

Panel beaters: cutting costs and generating energy

- The PV panels have generated 52,355 KWh of energy generation from 19th June 2013 to date, which is enough to power 11 houses for a year
- Received £4,303.08 up to 11th March 2015
- This initiative together with other Environmental impact mitigation actions positively contributed towards our Client being awarded ISO 14001 Environmental Management accreditation achieved in March 2015
- Helped achieve a positive BREEAM (Building Research Establishment Environmental Assessment Methodology) assessment on the building project. The final certification was gained in January 2013
- Demonstrated to our tenants environmental best practice and reduced energy costs and carbon footprint
- Our colleagues at Royal Priors Shopping Centre, Leamington Spa, subsequently instructed JoJu to install solar PV panels on a more recent building project for a Carluccio's restaurant
- LaSalle Investment Management are currently exploring an initiative to install roof PV panels across their entire property portfolio.

WHAT DID THE PROJECT INVOLVE?

Installation of 17.1kWh PV solar panel units, fixing accessories to the roof area and associated logistics –

electricals, distribution of wiring, testing and commissioning. Maintenance included checking voltages/currents at inverters and wiring and panels. The project cost approximately £47k which was paid for by the client's capital contribution to the building project. .

We originally estimated including 10% for shading 13,392 kWh/yr = 783kWh/kW. This was shown on MCS (Microgeneration Certification Scheme) certificate. In 2014 it generated 16,262 kWh (951kWh/kW/yr), which is 5% up on 2013 due to good weather, and in line with other systems. The new MCS guide using newer climatology data gives 872 kWh/kW.

The main advantage of solar power is that no pollution is created in the process of generating electricity. It is the most clean and green energy environmentally. Solar Energy is clean, renewable (unlike gas, oil and coal) and sustainable, helping to protect our environment.

WHO AND WHAT HAS BENEFITED?

The LIM/JLL Energy Reduction programme shows that High Chelmer has improved their energy usage over the last two years and are performing better than the Real Estate Environmental Benchmark (REEB).

PV panels were a first for High Chelmer generating electricity directly from the energy in daylight. They are a renewable source of electricity and emit no CO_2 during their operating lifetime. They are extremely robust and their performance is guaranteed for 25 years. Solar PV is incentivised under the Feed in Tarriff scheme, which rewards the user for every unit of green electricity generated.

CAN OTHER ORGANISATIONS BENEFIT?

Other organisations and communities are likely to benefit in the long term financially. Any positive environmental impact is one that all businesses should be encouraged to aspire to.

Rising energy costs are a major concern for any business. Government estimates predict a 43% increase in electricity prices and a 24% increase in gas prices, in real terms by 2050.

So installing a solar systems may help businesses counteract rising fuel costs. In addition, the system can be one of the most cost-effective ways to reduce the CO_2 emissions on new and existing buildings.

WHAT ARE THE LONGER TERM BENEFITS?

High Chelmer and its tenants have benefited from the ISO 14001 Environmental Management accreditation and high BREEAM rating for the building project. BREEAM is the Building Research Establishment's (BRE) Environmental Assessment Method first launched in the UK in 1990. It sets best practice standards for the environmental performance of buildings through design, specification, construction and operation.

We have demonstrated that we are an environmentally friendly shopping centre by the reduction of common area energy use which in turn has reduced our carbon footprint as well as costs. The following initiatives together with the introduction of the PV solar panels positively contributed to the above credentials:

- Waste Management – Our customer service team recycle tenants' cardboard and plastic waste as well as wood pallets. Our waste contractor Green Recycling have a policy of 100% recycling of our other waste
- Water Use – Internal supplies are metered as are external standpipes which are also locked secure when not in use. Rain water is harvested in two service yards which is used to irrigate external trees and planting areas
- Gas controls – radiators are all individually thermostatically controlled. Gas consumption was further reduced when water heaters at Threadneedle House were transferred to solar panels
- Electricity – All malls and service yards have the benefit of good (and free) natural light during the daylight period. Lighting is controlled via time clocks, photo-electric cells or movement detectors where applicable. Socket outlets in public areas are locked to prevent unauthorised use and for safety purposes. Power factoring devices are used in plant rooms. Utility service providers are centrally procured negotiating favourable tariffs. New lighting introduced during the recent refurbishment is extremely energy efficient. Under-utilised equipment is switched off when not in use, e.g. monitors at Customer Service Control are switched off overnight

as the single night officer uses the spot monitor only. This saves wear and tear on equipment as well as energy consumption. Movement sensor-lighting has been installed in common areas, such as management stairwells and corridors, as well as both staircases at Threadneedle House. The centre has an ongoing programme of conventional lighting replacement to Led

- PPM Programmes – regular maintenance regimes keep all plant and equipment at optimum efficiency. Access equipment is hired in block periods to co-ordinate high level cleaning and maintenance tasks
- Procurement procedures – lighting fixtures and fittings are purchased directly from a competitively tendered local service provider resulting in discounted costs benefit and avoidance of 'mark-up' by our electrical contractor
- Energy Performance Certificate (EPCs) surveys are carried out on all vacant units
- A lacewing habitat was planted and maintained at the rear of the building project
- High level bat boxes were installed at the rear of the building project
- High Chelmer has a fluvial and tidal flood plan registered with the Environment Agency
- High Chelmer are Gold members of Essex Wildlife Trust and have been for the last five years
- Programme of promotional environmental events including participation in the Edible Essex project.

High Chelmer entered the BCSC Gold Awards under the category of Refurbishment/Extension for the Primark project. We supported the application by hosting the judges' appraisal inspection and the compilation of a stop motion film piece which successfully highlighted the size and scale of the under-taking.

The client received a BCSC Gold Awards Commendation which was presented at the annual BCSC dinner in London on 5 December 2012.

High Chelmer also achieved a Secured by Design, the official UK Police flagship initiative supporting the principles of 'designing-out crime.' Secured by Design is owned by the Association of Chief Police Officers (ACPO) and has the support of the Home Office Crime Reduction & Community Safety Group and the Planning Section of The Office of the Deputy Prime Minister.

HISTORIC SCOTLAND

UPGRADING A TRADITIONAL COTTAGE

PROJECT AIM

The Gannochy Trust, in partnership with Historic Scotland, aimed to deliver a suite of energy efficiency measures on an inter war period traditionally built solid walled cottage as part of a wider sustainability review and stock upgrade.

Previous refurbishments had proved costly and resulted in technical issues with condensation as well as producing excessive quantities of waste and loss of durable building components. Key objectives for the project were:

- Revise the refurbishment measures to be more sustainable in terms of resources and waste, be more compatible for the building fabric, and give an improved indoor environment for tenants
- Evaluate the measures for effectiveness through in situ testing of U-values and relative humidity
- Develop a costed schedule of standard works that could be rolled out on other similar properties as part of a wider retrofit programme for the estate
- Upskill the local contractors and trades in the new range of measures
- Publish the works and share best practice in the retrofit of traditional and historic buildings.

PROJECT ACHIEVEMENTS

Measured thermal improvements to a solid wall property were achieved with materials technically compatible for a traditional building. The site phase went well with only modest adjustments to the schedule of works and no significant delays.

Good indoor air quality was ensured through the use of hygroscopic materials, improved conventional ventilation and the use of existing flues. Up-skilling and training in retrofit was delivered to local contractors and refurbishment techniques were refined and made more efficient.

Cottage industry: living with energy efficiency

The amenity and presentation of the cottage has been maintained through retention of linings and finishes. Damaged fittings were repaired and reused. Accurate cost control during the pilot allowed rates for the various measures to be developed - giving better financial data for planning the larger refurbishment programme. In situ u-value and humidity monitoring of the fabric has proved the effectiveness and durability of the measures.

Careful data input regarding the installed measures during the EPC process allowed an improvement in SAP rating. The client will adopt the new refurbishment principles, is pleased with the pilot work and now confident in the procurement and use of sustainable refurbishment materials.

WHAT DID THE PROJECT INVOLVE DOING AND WHY DID YOU DO IT?

1. This project involved the trial thermal upgrade of a three-bedroom traditional stone cottage, dating from the inter-

war period and owned and managed by the Gannochy Trust. This cottage, along with many similar properties, is located on the edge of Perth and are privately let to tenants at affordable levels.

Some of the stock had been improved between 2000 and 2014 using standard modern insulation techniques, but these refurbishments had proved expensive, disruptive and resulted in a lot of waste through the disposal of original materials, and much of the period charm of the interior of the cottages were lost. There were also technical issues with poor indoor air quality, condensation and mould.

2. As a result of this, a review of the refurbishment programme was taken by the Trust with input from Historic Scotland. The trust were keen to maintain the high quality of the internal and external amenity that was the vision of the founder of the trust who oversaw the development of the Cottages. This resulted in a new refurbishment strategy that emphasised the sustainability aims and objectives of the trust.

A new scope of works was developed retaining much of the fabric and making thermal improvements with materials technically appropriate for properties of this date and type. As many components were being retained there would be a significant reduction in waste. To achieve these reductions in waste and improved compatibility with the existing fabric a range refurbishment products were used in a new way.

These measures are largely new, with limited adoption in Scotland to date, most of which have been on other Historic Scotland Pilot projects. This revised scope of work gave a less invasive work programme, delivered at lower cost, with considerably reduced waste than the former refurbishment scheme. Key areas of the refurbishment were:

3. The methods used are replicable through a range of existing building types, especially those of traditional construction with solid stone walls, suspended timber floors and chimneys. The nature of the improvements mean that in most cases these measures would be suitable for Listed Buildings. To ensure that the experience gained in this project is available for use elsewhere the works have been written up as a Historic Scotland Refurbishment Case Study.

WHAT DID IT COST?

4. The funding for the works was provided internally by the Gannochy Trust and a Technical Research Grant from Historic Scotland of £25,000. The total works to the cottage cost £60,000, but much of this was standard upgrade items such as new kitchen, bathrooms and electrical wiring not covered in this submission.

WHAT ARE THE BENEFITS?

- Previously lowered ceilings in the kitchen and bathroom were removed resulting in higher internal air volumes and better air quality
- Blown insulation behind existing linings improved the thermal performance of the walls and ensured there was no loss of room space
- Retention and repair of lime plaster linings to walls and ceilings reduced costs and waste
- The use of clay paints in decoration contributed to humidity buffering and assisted in reducing the off gas during the re-decoration
- Lifting the suspended timber floors, insulating with wood fibre board, and relaying the original timber gave a significant improvement in thermal comfort, saved money and reduced waste
- An innovative infra-red heated panel, that doubled as a mirror, was installed in the bathroom. This device along with a heated towel rail, reduced condensation on the walls
- The roof space was insulated by the use of wood fibre board on the roof structure, allowing a warm storage area for use by the tenants
- Existing chimney flues were kept open for trickle ventilation in the bedrooms by the fitting of an adjustable grille. The passive measure contributed to a healthy internal environment and the maintenance of indoor air quality
- The existing internal doors, while damaged, were repaired and refastened with updated ironmongery. This allowed retention of character, and a significant reduction in waste.

The methods used are replicable though a range of existing building types,

ARE THERE LONGER TERM BENEFITS?

5. There was an emphasis on the simplicity and durability of the measures – all of which are expected to have a life which is essentially that of the building, and at least 50 years. The contractor who carried out the work needed a degree of instruction, but quickly appreciated the principles and bought into the new approach; this has generated capacity in the local area for similar work.

6. The reductions in energy use will be confirmed by comparison of actual energy bills (electricity and gas) before and after the works. This will happen in Summer 2015. The thermal performance of the walls and roof have likewise been measured by in situ measurements of heat flow, and pre and post intervention u- value readings obtained. These will be completed in Spring 2015. The EPC/SAP rating for the property has been improved from an E to a D.

7. As a result of this project, the trust now have a tested technical template for the improvement of the remaining stock; they have confirmed costs of each measure allowing better financial planning and they are improving their housing stock in advance of expected minimum energy efficiency standards for rented housing. Benefits to tenants include lower fuel bills, an improved indoor environment and a good aesthetic environment in which to live.

8. The success of the pilot means that the trust will now plan to roll out two or three of these refurbishments per year on similar buildings on their estate, confident that the measures work at a technical level, the costs are known, and tenants are pleased with the results. The works will also allow Historic Scotland to develop their publicly accessible guidance for the refurbishment of historic and traditional buildings in Scotland and elsewhere.

HYDRACHILL

DELIVERING THE FIRST FREE MAINS-FED DRINKING WATER SOLUTION TO UK STATIONS

PROJECT AIM

The Hydrachill project was born of frustration at the lack of access to hygienic mains-fed drinking water for 'on-the-go consumers', not least at public transport hubs where retailed bottled water options typically hold exclusive domain.

Conceived in 2007 under the Water for Health initiative, the project aimed to develop and deliver an attractive, durable water dispensing facility refillable to all containers, as a viable alternative to bottled water at high footfall locations.

Bottled water growth has raised serious environmental concerns, worldwide. In 1980 UK annual industry sales stood at 30 million litres. Today's figure is nearer three billion litres, 10,000 times greater and predominantly retailed in single-use bottles.

Yet UK mains-fed water quality is among the world's highest, deliverable by the most carbon efficient distribution system feasible. Each litre of bottled water generates a conservative 600 times the level of CO_2 than its tap water equivalent, emitting an estimated 350,000 tonnes of carbon into the UK environment annually. A mains-fed low carbon alternative is technically deliverable at almost any corresponding point of sale. HydraChill's aim is to provide and stimulate practical, hygienic, environmentally affordable choice in 'on the go' locations where bottled water profligacy has typically remained unencumbered.

On tap: free water at Liverpool University

PROJECT ACHIEVEMENTS

In 2014 Hammersmith became the UK's first station to offer free mains-fed drinking water for commuters, with the commissioning of a HydraChill Water Refilling Station. Hammersmith represented a significant project milestone for HydraChill. Satisfying such untried public-setting demands had required a lengthy prototyping, trialling and validating process at a range of suitable UK locations including universities, museums and sports-facilities. Following a prolonged process and with support from Thames Water and sanction from the Mayor of London's office, a custom-branded HydraChill machine was finally commissioned at TfL's Hammersmith Bust Station in June 2014.

Refillable to all containers and offering optional reusable bottles, the machine had immediate resonance. Within its first year Hammersmith's HydraChill has refilled more than 50,000 containers free, eradicated vicinity bottle litter and vended sufficient reusable bottles to sustain machine service and maintenance costs. An impeccable performance record has led to TfL both retaining the Hammersmith unit and commissioning two further machines at London River Services Tower and Westminster piers in 2015.

The Hammersmith project and pier installation response has shown that the public will readily choose a greener solution if available and that hygienic free water provision at commuter and other high footfall hubs is both feasible and viable.

Chilling out: free water at Westminster pier

WHAT DID THE PROJECT INVOLVE DOING?

The quality of UK mains-fed drinking water is acknowledged as among the world's safest, piped directly to our homes and buildings via a highly carbon-efficient method. However, for today's increasingly health and environmentally conscious UK consumer, the absence of convenient free and hygienic access to drinking water in public spaces is both a frustration and disappointment.

Adding urgency is the need to counteract the spiralling carbon emission impact of bottled water. Once considered by Britons as mainly a foreign travel necessity, shrewd marketing has seen bottled water consumption increase 10,000 fold in little over a generation.

Emissions consequence has been unavoidable. Producing one litre of bottled water generates 600 times more CO_2 than its tap water equivalent; nearly three billion UK litres are consumed, predominantly in litter and landfill-hungry disposable bottles. An issue causing worldwide reaction, NHC estimates 350,000 tonnes of CO_2 is released into the atmosphere every year from the UK bottled water industry alone.

The HydraChill project's origins lie with concept founder Nick Davis who in January 2007 was faced with a dilemma familiar to many business travellers and commuters. Having tap-filled his reusable bottle at home, the return journey railway stations offered no rehydration provision other than retailed bottle water purchase. Coincidentally his business meeting had been at the offices of Water UK where this frustration was shared and resolve expressed to devise a public access solution, promoting the quality and carbon energy benefits of UK tap water, and helping reduce disposable water bottle waste.

Subsequently, under Water UK's Water for Health initiative, a representative cross section of water utility companies provided guidance on key concept features and facilities. These should include customisable graphics, showcase visibility, public usage durability and refillability to all bottles. The machines should be industry accredited and critically avoid cross-contamination possibility via either bottle or human touch. The latter has been achieved with a retractable dispensing head and remains a unique HydraChill feature.

Further unique to HydraChill became the suggested phase 2 option of providing refillable water bottles from the unit, for those who might approach the facility without an available container. These could help reinforce the bottle reuse and retention message, while accrued revenue could help fund free water maintenance provision.

Initially it was hoped that quasi-government funding could be secured for prototype development. When backing was not forthcoming, Davis resolved to progress the development independently, conditional on water company machine placement support.

Agreement was reached with Welsh Water, Northumbrian Water and Thames Water for prototype installations in suitable trial locations. Hydrachill Ltd was established in January 2008 and sufficient private investment funding secured to meet the development and working capital needs of the business,.

After extensive development work, two phase1 water-only Hydrachill models were installed at Cardiff Metropolitan University in 2009. These were followed by prototype phase2 combined water and reusable bottle machines, installed at the Discovery Museum in Newcastle and Northumbria University's newly opened Sports Central complex (NUSC), the latter proving an invaluable testing ground.

WHO AND WHAT HAS BENEFITED?

HydraChill usage is recordable. The two NUSC machines commissioned in 2010 & 2011 respectively, will in late 2015 collectively reach one million 500ml water dispenses. Over nine thousand university-branded reusable water bottles have also been vended, supporting their upkeep. Effectively, just one UK building refilling up to one million bottles that might alternatively have contributed to bottled water litter, landfill and carbon emission output.

Transposing that impact across the UK by populating 'on the go' spaces with HydraChill or equivalently capable solutions and it becomes clear that the concept's CO_2 reduction potential is considerable.

It should be added that Cardiff Met have placed ten further 2-in-1 HydraChill machines around campus and Northumbria University four. Other universities have followed suit, again in almost all instances first machine placements being followed by additional installations. Student communities have shown a clear lead in adopting this win-win service for their personal and environmental wellbeing. With greater option awareness, other sectors could follow.

Despite this generic university progress, the founding objective of a public transport facility had yet to be fulfilled. In June 2014 Thames Water honoured a long-standing commitment to sponsor a machine under their Londonontap banner at Hammersmith Bus Station. Sanctioned by the mayor's office, machine and reusable bottle graphics encouraged users to Help Reduce London's Carbon Footprint.

The Hammersmith machine thus became the first free mains-fed drinking facility at any UK station. Hygienic, accessible and free to all irrespective of means, its commuters were first to no longer suffer the frustration experienced by HydraChill's founder in 2007 and represented a stage mission fulfilled.

Within its first year more than 50,000 users have become familiar with and utilised the Hammersmith service. TfL independently sanctioned two further machines, at London River Services Tower and Westminster piers. Their arrival was announced in London Evening Standard web and Facebook articles 16th June 2015, which became that week's most highly shared and 'liked' items, remaining so to date of writing. Comments make clear that the principle strikes a genuine public chord.

To help achieve equivalent CO_2 reductions nationally, HydraChill's challenge is to capture this goodwill, raising the project profile so that more communities could readily adopt this principle; transport, education, local authorities, conscientious major retailers etc.

After eight years commitment there is arguably a limit to what Hydrachill Ltd directors can achieve without further investment or support. Unarguably however the principle has simply demonstrated how the public free water objective can be achieved and popularly appreciated. Sponsorship will be key to progress, ideally combining media space potential and goodwill into an even more dynamic proposition, possibly with the support of a high-profile spokesperson.

HydraChill has attempted to revitalise the case for improved public hydration access in an innovative, appealing and more sustainable format. With assiduousness and support the project has the potential to make a positive impact, help raise public consciousness of important environmental and health issues and solutions and stimulate a genuine counteractive momentum for valuable carbon emission reduction.

HYDRACHILL WATER REFILLING STATIONS

A vibrant alternative to the 'environmental insanity' of disposable water bottles

In barely 5 years, a university department in Newcastle, England has saved over one million plastic water bottles from litter and landfill, reducing equivalent carbon emissions to a fraction.

Northumbria University Sports Central invested in two customised mains-fed HydraChill Water Refilling Stations, hygienically refillable to all bottles.

HydraChill has been purposely developed to provide reliable access to public drinking water, in an attractive message-brandable format at key 'on-the-go' locations.

HydraChill's '2-in-1' machines also offer unique optional access to refillable water bottles. Free water can be supported from reusable bottle sales, or sponsored advertising.

If one building in one city can eliminate one million energy hungry bottles from disposal, imagine the cumulative effect were this practice adopted as public principle?

Others have now followed Northumbria's example. HydraChill has become the first free drinking water service for Transport for London commuters at Hammersmith and London River Services.

HydraChill. Working to improve public access to drinking water. Proving how imaginative solutions can have consumer appeal.

WWW.HYDRACHILL.CO.UK HYDRACHILL

J MURPHY & SONS

WHITE WALTHAM ECO-GARDEN

PROJECT AIM

The decline of the honey bee is a hot environmental topic. The company had a campaign regarding planting sunflower seeds in order to help the honey bee and highlight the problem.

It was decided to construct an Eco-garden within the White Waltham depot by using waste material around the site compound. The aim was to improve the visual appearance of the depot and provide opportunities to enhance the biodiversity of the area.

PROJECT ACHIEVEMENTS

- We successfully managed to construct the garden by reusing waste material around the yard. Old pallets were used for the fencing and faulty concrete pipes were used for the flower beds. Bird houses were constructed from timber waste.
- Sunflowers planted around the garden have attracted an abundance of insects to the area. Bumble bees are a common feature this year. Our sunflowers have been recorded at heights of more than two metres
- A healthy bee hive was installed in the garden. During the spring the beehive produced a swarm which was happily taken by the local beekeepers. We continue to look after the hive and began harvesting the honey this year.
- Birds have nested within the gardens bird houses. and chicks have been observed flying from the nest
- The overall visual appearance of the yard has significantly improved. Workers at the site enjoy the area as a place to relax when the weather is clear.

WHAT DID THE PROJECT INVOLVE DOING AND WHY DID YOU DO IT?

During Climate Week the company had a campaign highlighting the decline of the honey bee. The environmental department were encouraging workers to plant sunflowers seeds to highlight the issue.

Flower power: sunflowers add a splash of colour

At the White Waltham Depot we decided to improve the appearance of a section of the compound with an Eco-garden. The garden would be complete with an active beehive. The aim was to encourage other projects of its kind within our business and improve the biodiversity of the area.

Plans were drawn up for the Eco-garden. It was decided to reuse as much waste material around the yard as possible. Old Timber pallets, paint cans, failed concrete pipes, timber waste provided some of the construction materials. A team of five volunteers was involved in the construction of the yard. The failed concrete pipes were turned into flower beds within the middle of the yard.

Timber pallets provided the boundary fencing material. Excess aggregate provided material for the path. Sunflowers, honey suckle and other flowers were purchased from a local supplier.

WHAT DID IT COST AND WHERE DID THE MONEY COME FROM?

This helped to greatly reduce the costs of the project. The main cost was attributed to the purchase of the bee hive (£400.00) with very little spent elsewhere.

The beehive was locally sourced and funds were provided by the company's environmental department. A bird house, bird bath and feeding station were also installed.

Team work: the yard has been transformed into a garden

When the garden was complete, the bees were brought to the hive from a local bee keeper. The bees quickly took to their new habitat and surrounding environment. They began producing honey very quickly. We started harvesting the honey earlier this year with a number of jars donated to our workforce.

The local bee keeper has provided training and tips for looking after the bee hive. The sunflowers in the garden have been recorded of heights up to two metres and attract an abundance of insects.

Bird nesting with the garden is very evident during the spring and early summer. Workers have witnessed young chicks fleeing the bird house.

WHO HAS BENEFITED?

The overall visual appearance of the yard has been greatly improved. The garden provides an area for workers to relax and enjoy nature in what was once a dull material storage area. The project was published as a good practice by the company. A case study on the project was communicated via email to all construction managers.

Minimal effort is required for maintenance. Throughout the business opportunities are explored to enhance the environment and reuse waste material.

JONES LANG LASALLE (JLL)

ENVIRONMENTAL IMPROVEMENT: WESTSIDE PLAZA SHOPPING CENTRE

PROJECT AIM

The Westside Plaza Shopping Centre is proactive in its approach to the management of environmental issues. Whenever reasonably practicable we will introduce initiatives, technology and policies to demonstrate our commitment to promoting environmental benefits.

We are dedicated to managing the centre's operations to reduce energy, water and waste. To this end we will regularly review our environmental management to ensure continual improvement and maintenance so that arrangements operate effectively.

Our aim is to make certain that regulatory requirements are consistently met, to increase operational efficiency, minimize waste and develop relationship- building with our tenants, employees and customers to work in partnership towards environmental improvement.

PROJECT ACHIEVEMENTS

We are a community based shopping centre with a high percentage of unemployment and are very financially restricted. We have achieved excellent results in the shopping centre especially in the current retail climate. The cleaning staff achieved a Gold & Silver Award from the local council in keeping Edinburgh clean.

Our ethos is that the general amenities and physical environment should promote a feeling of safety and good order to help create, safe interaction within the centre. In today's market this will not only help enhance the centre's reputation but will also encourage return visits.

In this ever-changing world, with new rules and regulations affecting the way we operate and where litigation seems to be the normal course of action should we fail to comply, these and other factors can have a serious effect on our shopping centre.

As a company, we try to raise awareness of these factors that will help to protect our business interests by reducing the chance of litigation through negligence on our part. The level of accidents and incidents has dropped dramatically through this proactive approach.

I am interested in providing training to our cleaning team to a recognized standard like Green Apple and for the level of skills within the cleaning team to also be recognised. I think this would provide a feel good factor and may also be financially rewarding for the cleaning team.

WHAT DID THE PROJECT INVOLVE DOING AND WHY DID YOU DO IT?

- Working in co-operation with Edinburgh City Council & Environment Waste Control Ltd in our endeavors to keep the City in a clean attractive environment for both its tourism trade and citizens
- To control and recycle waste products to reduce landfill in line with current legislation and environmental control.

As a large community business we had a duty to our client, tenants and customers to provide a safe and secure environment. The cost was minimal and was taken from the service charge budget.

By reducing the landfill requirement this reduced the annual cost to the client and brought in additional income for the recycles waste, cardboard and plastic.

By reducing the landfill this would bring the centre in line with the future proposed requirement by have a negative landfill requirement and protect the environment .

If we all do what is necessary, even so small, we can save the environment for future generations. If all businesses were to conduct this type of venture, the amount of waste being reduced to landfill would be significant.

KELTBRAY

BBC TELEVISION CENTRE

PROJECT AIM

The aim of this project was to improve management and control of waste and resource use impacting the environment from Keltbray operations by implementing new environmental reporting software.

To fulfil this requirement Keltbray purchased the online software SMARTWaste, which measures waste and resource use and enables each site or project to benchmark against project, company and industry performance and targets.

It also allows live comparison of actual versus forecast waste from a pre-demolition audit. The other reporting elements in addition to waste are; water, energy, timber and fuel use as well as carbon footprint.

The intention is to adopt a uniform reporting method to enable us to better compare sites environmental impacts and enable us to make more accurate predictions of similar new sites impacts. This will aid us in setting of targets that are both ambitious but attainable for each individual project.

The overall aim is to implement this software across all Keltbray business units to ensure best practice waste and environmental management company wide.

PROJECT ACHIEVEMENTS

SMARTWaste was chosen by Keltbray as the right software for streamlining environmental reporting of sites and to bring together the measuring and monitoring of all environmental impacts in one easily accessible database.

The software has and will continue to assist development and setting of both project specific and business targets to reduce waste and resource use and also helps us to identify our largest impacts for targeting.

The software live tracks actual progress against these targets and has helped sites overall waste management by ensuring that the waste hierarchy is applied at all operation levels to favour prevention, reuse and recycling of materials generated from the demolition projects or site activities.

SMARTWaste also has helped us improve our Duty of Care checks as it keeps an accessible database of all approved waste carriers and facilities licences and permits, and highlighting any that are close to expiring.

This provides an automated solution to our current process of keeping licenses and permits up to date and prompts us to seek updated licenses when necessary. Time and effort savings have been made for project employees in relation to environmental reporting as previous reporting was more manual and labour-intensive.

WHAT DID THE PROJECT INVOLVE DOING AND WHY DID YOU DO IT?

The project involves the demolition and associated works at the BBC Television Centre at White City in West London. The building opened in 1960 and was home to BBC television until 2013. Keltbray are working as one of two contractors for the demolition works under principle contractor MACE.

The construction period the television centre was built in also means that materials containing asbestos are present throughout. Keltbray Environmental Solutions Ltd is being used as the contractor for all asbestos removal activities and due to the hazardous nature of the asbestos waste, careful management and recording is needed.

The nature of demolition gives rise to large quantities of material classed as 'waste' that need to be effectively managed and removed from site and the process recorded at every stage. Fuel needed for plant, timber used for temporary purposes and water and electricity use for the temporary sites all have an environmental impact that businesses have an increasing responsibility to report and reduce.

SMARTWaste was therefore trialled and selected as an appropriate tool to effectively record environmental impacts and communicate to key parties. The tool efficiently tracks waste quantities and types as well as the responsible parties for transporting and receiving the waste removed from site.

The data entry fits in with previous site methods of recording waste movements for other purposes, such as for accounts, so has integrated in well to existing site practices.

It is the policy of Keltbray Group to make a positive contribution to the world that we live and work in, and be the best in its sector at everything the company does. For Keltbray, this means that all construction projects need to maintain and continually improve recording, reporting and reducing of waste and resource use related to the company's operations.

For this reason SMARTWaste was selected and implemented as the approved online reporting tool for the BBC Television Centre Project, which is now being implemented across all other business divisions. Waste reduction and improving resource efficiency are key considerations to the way we work and we aim to balance our economic achievements with our contribution to society and the environment.

WHAT DID IT COST?

The money for purchasing SMARTWaste was sourced from reusing office equipment in other projects and by savings on waste transport journeys. The demolition arisings from the project was crushed and re-used on site as a piling mat.

WHO AND WHAT HAS BENEFITED?

Keltbray Ltd, our clients and employees have benefitted from the implementation of SMARTWaste software, as the tool reduces the effort and time required for reporting for Keltbray personnel and allows data to be easily shared with principle contractors and the client for their own reporting purposes.

We are able to now accurately measure and reduce our resource use and use most favourable waste management routes for waste arising from our activities, not only at the BBC Television Centre project but also across all business divisions. In the long term this will help us to achieve our target of minimising waste and ensuring we supply zero waste to landfill by 2020.

SMARTWaste for this project is fully integrated with other sustainability management modules such as FSC Timber Certification, Carbon emissions, pre-commencement waste quantity audits enabling us to set up an overall target as well as specific.

CAN OTHER ORGANISATIONS BENEFIT?

Other organisations and communities who purchase and implement the software we use will benefit by improved efficiency of environmental reporting of waste, water, energy, fuel and timber use.

WHAT DID YOU LEARN?

The main learning outcomes to date from implementing SMARTWaste were that we are able to monitor environmental performance and assess the phases of projects that have the most significant impacts for each environmental factor. For example the main uses of timber are at the start of the project for building of hoarding, so re-use schemes can be planned in the pre-project stages alongside the ongoing sourcing of sustainable timber (FSC certified only).

The help targeting areas with the most environmental impact will contribute to the overall sustainability performance of new developments.

We plan to implement the software across the group in order to accurately record and reduce our waste and resource use for other divisions such as Keltbray Rail, Plant, Remediation and Environmental Solutions (asbestos removal).

KELTBRAY

VICTORIA CIRCLE

PROJECT AIM

The overall aim of the project was to reduce air pollution emissions and congestion on already busy surrounding roads by removing the need for a large amount of waste removal and material delivery transport to and from the Victoria Circle project, which was located on some of the busiest streets in central London.

In order to do this the Victoria Circle team proposed to set up a materials management and logistics plan in order to crush all concrete arisings from the demolition works onsite for reuse and recycling. It was thought that this was the best practice method for sourcing the large quantities of material needed for backfilling of basements and construction of piling mats.

By retaining the material on site, the need for the material to be removed as waste from the demolition stages, and so creating a future requirement to import crushed concrete transported in from other sites for future work stages was eliminated.

This was aimed to keep air pollution from vehicles associated with the project works to a minimum, to benefit public health as well as the environment and also ease the impact on local congestion.

PROJECT ACHIEVEMENTS

The Victoria Circle project successfully managed 53,000 tonnes of demolition material for reuse in latter project stages. One of the greatest accomplishments was the successful management in moving of stockpiled material around site to facilitate the changing site conditions and works. This materials management was on a large scale and we regard it as a great success to have efficiently completed the work on a Central London site.

The works removed the need for waste removal and material delivery transport for site and avoided 5,400 vehicle movements and the associated carbon emissions from the local area. This in turn helped us to reduce the projects carbon footprint and improve the overall environmental performance. Another great achievement was the challenging organisation of logistics at the end phases of the project, where all the reused and additional material from the dig works needed to be removed and therefore the site was managing 120 haulage lorries in and out of site per day.

The main challenges overcome were both the logistics within the site as well as outside of site, as the project was located on some of the busiest streets in London.

WHAT DID THE PROJECT INVOLVE DOING AND WHY DID YOU DO IT?

The Victoria Circle project involved the demolition of 22 different buildings down to ground level followed by the installation of temporary works, demolition of ground slab, piling enabling works (including backfilling of existing basements) and a double storey, top down dig.

The surrounding area comprised commercial buildings including offices, hotels and theatre facilities, retail units, residential and Victoria Station Upgrade construction sites. The streets around the perimeter of the site are some of the busiest in London, both in terms of public footfall and commuter/tourist road traffic. Minimising the impact of lorries on the road was key to ensure the safety of the general public and to reduce associated carbon emissions as far as possible.

The nature of demolition works creates great quantities of materials classed as 'waste', some of which can be directly re-used on site but ultimately need to be efficiently managed and removed from site.

Once demolition of the buildings had been completed at the Victoria Circle project, a large quantity of crushed

Model rescue: the developers made sure that the Allingham Horse sculptures were saved intact

concrete was needed for the backfilling of basements and construction of piling mats.

As 'waste' concrete was abundant in the structural demolition phase of the project, it was decided that this concrete would be crushed and managed on-site to prevent it being removed as waste and so creating a future requirement to import crushed concrete transported in from other sites.

Despite the thorough planning and ongoing management needed to systematically move the stockpiled material around site to facilitate the further works, retaining the material for re-use was seen as the best practice option in terms of reducing the amount of vehicles on the busy roads and associated emissions.

There was also a significant cost incentive in avoiding the removal of the existing materials followed by purchase and import of backfill and piling mat materials.

Reducing carbon and other emissions from all our projects is a key consideration both in terms of lessening local pollution from vehicles, for the better health of those surrounding our sites and looking at the wider global picture, to play our part in action against climate change.

There is increasing pressure from government to lower emissions especially in London where most of our projects are concentrated. The introduction of the London Low Emission Zone (LEZ) is one of the driving pressures to keep haulage trucks needed for waste transportation to a minimum.

The traffic and materials management for the works was extremely difficult and an important part of delivering this job safely and on time. This was successfully achieved through a very high level of daily coordination, which was fundamental because of the complex logistics changes that were needed to facilitate the differing demolition phases, the multiple services removals, the hoarding and pile probing works.

The site entrances and exits frequently changed position and stockpiles of crushed concrete needed to be methodically moved around the site as the project evolved.

The number one priority was the safety of the thousands of pedestrians, cyclists and other road users around the development, who are constantly passing and interacting with the site entrances at all times, for this reason also it was beneficial to reduce the number of vehicles travelling to site.

In total 24,000m3 of demolition arising's were maintained on site, the equivalent to approximately 2,700 lorry movements off-site and a further 2,700 movements bringing material back to site. Once the piling works were complete 2,700 loads of material were removed from site, meaning that overall the recycling and re-use of materials on-site avoided 5,400 vehicle movements and associated carbon emissions.

The cost for the reuse and associated management of stockpiled material did not bear a cost in monetary terms. There was a significant cost benefit from avoiding removal of existing materials and then needing to purchase and import of new material to use for backfill and piling mats. The costs were instead seen in time and effort needed for the ongoing planning and implementation of the management of materials.

WHO AND WHAT HAS BENEFITED?

The main beneficiaries from onsite material reuse and accompanying reduction in local vehicle traffic were; Keltbray who saw a cost benefit in addition to helping to reach company and project carbon reduction aims, the general public from eased traffic congestion in an already busy location and preventing an increase in localised air pollution from the avoided 5,400 vehicle movements and lastly, in the longer term the local authorities benefited by the reduction in potential emissions to air from our works so as not to put additional stress on the Westminster Air Quality Management area.

A campaign was initiated to save the The Allington House Sculptures on the project, which gained a lot of support. Keltbray used innovative techniques to remove the sculptures undamaged and preserving them, providing the logistics free of charge. A crane was used to lift a specially created steel cradle holding the beam that the sculptures were mounted upon.

CAN OTHER ORGANISATIONS BENEFIT?

Other organisations and communities who apply the same materials re-use and recycling method can benefit from cost savings and improvements in local emissions from related transport reduction. The method could be replicated on many different types of projects and materials.

WHAT DID YOU LEARN?

The main learnings from the project were those gained in the planning, ongoing organisation and skills to manage the materials around site. In regards to further development, the success of effectively managing materials on a large scale, on a logistically challenging London site will enable us to use this technique on other projects.

The project won the Urban Demolition Award at the World Demolition Awards in 2013 and gained 'Excellent' under the Considerate Constructors scheme.

KIER

HI-VISIBILITY RECYCLING SCHEME IN MOLDOVA

PROJECT AIM

- Diversion of unwanted hi-visibility clothing from a company buy-out to a road safety program in Eastern Europe, where improvements in infrastructure have put lives of local residents at risk
- Setting up an alliance of organisations to help collect, deliver, clean and distribute the items and to further develop these relationships to create additional opportunities to help road safety in this region of Europe.

Visionaries: donations will help improve road safety in Moldova

PROJECT ACHIEVEMENTS

The project has been a resounding success. The donation of approximately 5,000 items of hi-visibility cloth has not only helped improve road safety in Moldova but has also eliminated quality usable equipment from needlessly going to landfill sites in the UK.

The project has also sparked the idea and trial of a secondary project sending unwanted but usable child car seats to impoverished areas in Eastern Europe. The project has also created strong links between the various project partners. It is hoped that this is just the start of many partnership projects to come.

WHAT DID THE PROJECT INVOLVE DOING AND WHY DID YOU DO IT?

For more than 100 years May Gurney had been a civil engineering and services business based in Norwich, but in 2013 the company was bought by a larger competitor, Kier. With over 12,000 staff based from Scotland to Cornwall, the integration of the two businesses was going to take time.

It was decided at an early stage that the May Gurney brand, where contractually allowed, would be changed to Kier. This meant that corporate branding across almost the whole business needed to be changed, which included livery on thousands of vehicles, as well as on thousands of items of branded clothing.

Jamie Edmonds, May Gurney's Group Environment Manager, recognised that useable items of branded clothing both new and used were going to be thrown away, so he set out on a mission to find a good use for the unwanted items.

Through intensive research, contact was made with Emma MacLennan, Director of he Eastern Alliance for Safe & Sustainable Transport (EASST), who saw a huge potential for the hi-visibility clothing to be put to good use in Moldova as part of the charity's road safety projects.

EASST is an independent UK-registered charity working in Eastern Europe, Central Asia and the South Caucasus, helping to facilitate cross-border projects on sustainable transport and road safety. The charity works through local partner organisations to make their efforts more effective by sharing know-how and resources to save lives and prevent injuries on the roads.

EASST's work in Moldova began in 2009 with the launch of a Make Roads Safe Campaign with the Moldovan non-government organisation, the Automobile Club of Moldova (ACM). This led to the creation of a new National Road Safety Council chaired by the Moldovan Prime Minister and involving key ministers and the traffic police.

It was decided by EASST and ACM that the hi-visibility clothing would be best donated to traditional farmers from rural villages, who still use horse carts to move produce and equipment, often having to make long journeys on fast highways.

Differential speeds between motor vehicles and the horse carts have been putting farmers and other road users at great risk, especially at night during the winter months.

WHO AND WHAT HAS BENEFITED?

Over an 18 month period approximately 5,000 items of hi-visibility clothing were collected from across the former May Gurney business and the wider Kier Group. Even suppliers Anchor Safety, who had pre-printed hi- visibility vests, donated their holding stock of 200 vests, as they thought the project had such merit.

Getting the clothing to farmers in Moldova involved a chain of organisations, each playing a specific role in the delivery. Once the clothing had been collected from around the UK to a central collection point by Kier, it was transported to Eastern Europe with assistance from the US embassy in Moldova and is now being distributed as part of an embassy-aided project supporting community policing.

ACM facilitated the donation to the Moldovan Road Police and helped to identify areas where farmers with horse carts were at greater risk. The first villages to receive the hi-visibility clothing were those where EASST and ACM conducted road safety training for civil society organisations and community leaders.

This was part of a European Bank for Reconstruction and Development (EBRD) technical cooperation programme, targeting villages exposed to major road rehabilitation schemes. The communities themselves identified a need for help with equipment to improve the visibility of agricultural vehicles on the roads.

The first of the hi-visibility clothing distributed was given to farmers in the village of Lăpuşna in the Hincesti region, situated 54km southwest of the capital Chişinău. A short ceremony was held at the village school, where the clothing was officially handed over to local farmers and local police.

Delegates from the US embassy, EASST, Kier, ACM and the Moldova Police attended the ceremony and helped hand out the clothing to the local farmers. Further distribution of the clothing will be completed by ACM and the local police.

WHAT ARE THE LONGER TERM BENEFITS?

The project has been a resounding success. The donation of approximately 5,000 items of hi-visibility cloth has not only helped improve road safety in Moldova but has also eliminated good quality useable equipment needlessly going to landfill sites in the UK.

Kier is continuing its in-house collection of hi-visibility clothing and is also now collecting suitable clothing from its working partners and competitors. Once sufficient quantities have been gathered, a second consignment will be sent to Moldova for further distribution.

ARE YOU PLANNING FURTHER DEVELOPMENTS?

During the ceremony in Lăpuşna, discussions were held between EASST, Kier, ACM and the US Embassy. It was recognised that other road safety equipment could be donated from Kier

An easy and obvious place to start was with the collection of unwanted childrens' car seats at Kier's household waste recycling centres (HWRC). Due to restrictions in the UK,

thousands of good quality car seats are disposed of to landfill every year, rather than being reused or sold on.

A trial collection of the seats will take place at Kier's North Somerset contract. Collection, cleaning and thorough safety checks will be completed by another charity partner, Groundwork South, before the car seats are sent to Moldova for reuse in poor areas where they are too expensive to buy. The AA has been consulted, advising on how to ensure the second-hand seats are not second-rate.

The trial will continue throughout 2015, with the aim of rolling out the programme across all Kier's HWRC sites in 2016. This could result in thousands of car seats being saved from landfill and reused every year by needy families.

The project has also created strong links between the various project partners. It is hoped that this is just the start of many partnership projects to come.

KIER CONSTRUCTION

CARTERHATCH LANE

PROJECT AIM

Innovation and best practice are at the core of Kier's commitment to the environment. We want to be recognised as one of the leaders in our field and we always seek to introduce innovative solutions which will benefit the environment while reducing costs and inefficiencies.

Kier Construction London looked to their supply chain for support on achieving these innovative solutions. Through discussions with our supply chain at a regional level, three different Schemes were seen to have the potential in achieving environmental benefits to our business.

The Packaging Return Initiative, Community Wood Recycling and HaziBag Schemes offered exciting opportunities for Kier to reduce waste to landfill (Kier target of 5% year on year), reuse materials, reduce emissions, segregate waste efficiently and save money.

Rather than rolling initiatives out to all sites, Kier focused on Carterhatch Lane as an opportunity to trial the schemes together. Kier are constructing 150 new affordable houses, maisonettes and apartments in Carterhatch Lane, Enfield, which are being built to Level 4 of the Code for Sustainable Homes.

Carterhatch Lane is a Timber frame construction and is already seeing the benefits of reduction in waste from off-site construction, Kier wanted to go further when it came to materials being delivered to site that generated waste.

Kier & Travis Perkins partnered together to introduce a Packaging Return Initiative, which involved the collection of cardboard packaging, plastic wrapping and wooden pallets on delivery of products to the site. The initiative aimed to reduce waste to landfill, save money and reduce vehicle movements on site.

The Packaging Return Initiative involved Travis Perkins issuing Kier with bulk bags for cardboard and plastic segregation. The team built a site storage facility for holding the bulk bags and used a baler to maximise space in bags for collection. The procedure was as follows:

- Travis Perkins will collect cardboard, plastic wrapping and pallets on delivery of materials
- To arrange collection state number of bags for collection, when placing material orders
- To obtain replacement recycling bulk bags and place request at time of order.

Kier & Community Wood Recycling was introduced to Kier Construction London as an alternative to disposing of wood waste in a mixed waste skip with it ending up going to landfill.

The Carterhatch Lane team were committed to keeping their environmental impact to a minimum. Part of that was finding the most sustainable way of dealing with waste. As a company we are always looking to act more sustainably and by adopting the best environmental practises that are available, we hope to be a beacon of best practises for the industry- encouraging other firms to take part too.

This meant that a change in working practises on a busy, space-hungry site. The Carterhatch Lane site took it upon themselves to embrace the scheme and to segregate an area where all wood waste could be saved for collection and re-use by CWR. To help ensure the success of the scheme, the site spent lots of time and effort encouraging all workers to take part so the whole site could see wood as a resource rather than a waste.

Kier Construction and London & Windsor Waste Management Ltd teamed up for the first time together to use Hazibags, keeping with Kier Group's commitment to minimising environmental impact.

The UN approved Hazibags were adopted at the start of the development and the flexible intermediate bulk container was used on site for potential spills and to store COSHH waste including mastic tubes, sealants and paint tins. This provided a solution that improves the management of hazardous waste collection, segregation and disposal benefiting the environment by carbon emissions and transport reducing costs.

WHO AND WHAT HAS BENEFITED?

Scheme 1 (Take Back Initiative)

The scheme has not only saved the project money, as over 400 pallets have been collected rather than put in skips, but it has also allowed the pallets to be reused rather than recycled.

There has been over 98 cubic metres of cardboard/plastic packaging collected which would have otherwise been put in a skip. This has saved the equivalent of around 16 eight yard skips .

The further benefits of this scheme are based around "reverse logistics". Travis Perkins makes frequent deliveries to the site and collects the recycling and pallets when deliveries are made. This avoids the need for a specific collection service which has avoided extra vehicle movements and associated emissions.

With an average truck taking away 30 pallets at a time, that has saved around 13 individual collections to the site with the pallets being collected at same time of material delivery reducing a lot of carbon emissions.

Wood work: recycling has led to job creation

Scheme 2 (Community Wood Recycling)

There has been a significant reduction in waste disposal costs on site. More than 336 cubic yards (33.60 tonnes) was collected from site. This costs a total of £5,376.00 – a saving of £1,155 compared to using a normal 12 yard skip.

Further benefits include less waste to landfill, reduced CO_2 emissions through timber reuse, local communities having a source of low-cost quality timber and disadvantaged people having job opportunities created in the community wood recycling by using the scheme.

In the bag: Hazibags take up far less space

Scheme 3 (HaziBag)

As empty bags are flat-packed and take up far less space in storage than traditional containers they help to keep the site tidy, improving general housekeeping and the health and safety site staff. Flat-packed Hazibags take up far less space in delivery than traditional containers. More than 100 flat-packed Hazibags fit on a pallet taking up the same space as four x 200 litre drums,

Through the success of these initiatives, Kier have rolled schemes across other sites and eventually plan to capture all sites.

WHAT ARE THE LONGER TERM BENEFITS?

Community Wood Recycling

Whilst we continue to roll this scheme out, to date the Community Wood Recycling project has rescued a total of 9,120 cubic yards (912 tonnes) of wood from our sites. This has directly led to the creation of around ten permanent jobs. In addition, 61 local "disadvantaged" people have benefited from an average19 days of useful training. With the scheme now in operation on more Kier sites, the positive outcomes will only multiply bringing further benefit to the environment, local people and society in general.

Hazi Bags

By rolling the product out across all sites we are continuing to educate our staff and subcontractors about the importance of segregating and disposing of hazardous waste compliantly. We are further improving site safety and good housekeeping as well reducing our carbon footprint with every site that uses the Hazi-bag.

Packaging Return Initiative

This scheme really got the Carterhatch team thinking about applying the "waste hierarchy" to avoid the generation of waste through waste segregation on site to maximise reuse and recycling. This will influence other projects to take on a similar approach with waste which would have traditionally been dealt with by placing pallets and packaging in a mixed general waste skip.

Whilst other companies can incorporate these schemes into their own best practice business models on site or in the office, the initiatives are only successful through collaboration between the procurement team, supply chain and site team.

The schemes need to be managed, communicated and have support from all parties involved.

KIER HARLOW

THE COMMUNITY WAVE

PROJECT AIM

To improve the quality of life for the residents of Harlow by reducing the National Indicator (NI) 195 score for litter and detritus (waste material or rubbish) by the smarter use of existing resources and with a background of decreased local authority budgets.

NI 195 is the street cleanliness indicator, which measures levels of litter, detritus, graffiti and fly-posting on our streets and in our neighbourhoods. It is scored by Keep Britain Tidy. The visits are conducted four times a year but are varied to avoid contrived results.

PROJECT ACHIEVEMENTS

The Streetscene service had a score of 21.95% for detritus and when the new way of delivering the service was introduced, a target of 4.23% was set.

The street cleaning service recorded a score of 2.98% during the last visit from Keep Britain Tidy. Mick Jenkins, Service Manager, Environmental and Waste, Kier Harlow, said: "Over the past year we have been addressing litter and detritus issues around Harlow Town. Now that the results are in, I am very pleased with the efforts of our street cleansing teams."

Clean up: a tidier Harlow

WHAT DID THE PROJECT INVOLVE DOING?

In 2014 more than 11,000 tonnes of litter and detritus were removed from the streets of Harlow, with much of the waste being recycled or composted. This included 200 tonnes of leaves sent for composting and 300 tonnes of paper collected for recycling.

The key to the project's success was to analyse existing working patterns and change the way the service was provided. Project Wave ensures that the town is swept clean on a seven week cycle, which is down from the more traditional 16 weeks.

There is more communication between other service providers in the area. For example, if the grass cutting teams are in a certain area, the Wave ensures the street cleaning teams follow on to sweep up the cuttings.

There is a team of 19 people which includes 14 street sweepers and five community beat sweepers. Two new mechanical road sweepers have been introduced to supplement the people on the ground.

The community beat sweepers target problem areas in the town and provide a more intensive two week cycle in these hot spots.

The street teams also keep an eye out for people who are sleeping rough and pass the information on to the council's homeless team.

The teams have mobile phones so they can contact the enforcement team direct if they see fly-tipping or abandoned cars to allow them to investigate before the rubbish is removed. There is a dedicated community enforcement officer who works for the local authority but is essential to making the partnership between Kier and Harlow Council work.

Street cleaning complaints have reduced by 50% and customer satisfaction was at a high of 83.33% in December 2014, up 6% since the Wave was introduced.

WHY DID YOU DO IT?

The project was undertaken to greatly improve the performance and service being provided to the citizens of Harlow and clean up by bringing down the town's NI 195 score.

WHO AND WHAT HAS BENEFITED?

The residents have a more pleasant town in which to live and the workforce's morale and feeling of job satisfaction has increased. 90% of the workforce live in the areas and take pride in the success of the project.

Other opportunities have opened up for the street teams with individuals being trained up to work in the office on Wave logistics. There is less road flooding because road gullies are not a cutting blocked causing surface water ponding. And clean streets improve public health with a reduction in rodent populations.

WHAT DID IT COST?

Staff costs were neutral as the costs associated with the mechanical road sweepers and the community enforcement officer, which is approximately £95k per year, came out of existing budgets and therefore straight off the bottom line (profit).

CAN OTHER ORGANISATIONS AND COMMUNITIES BENEFIT?

It could be duplicated in any local authority by working in partnership with their chosen contractors and delivering it as a joint operation.

WHAT DID YOU LEARN AND ARE YOU PLANNING ANY FUTURE DEVELOPMENT?

We are exploring introducing the key principles into our other street cleaning services – in Bromley for example.

KM CHARITY TEAM

KENT GREEN TRAVEL CAMPAIGN

PROJECT AIM

To encourage families and schools to regard green travel to school as a practical and normal behaviour – and to ease them off the school run. The aim is to reduce traffic congestion at the school gates, reduce pollution levels, and provide regular exercise for children and families and to teach children vital road safety skills.

The 'business side' of the charity's work has been to fine-tune existing schemes of Active Bug, Walk on Wednesday and Walking Bus while developing the new more portable Green Footsteps initiative. And to develop a new cloud-based data capture area for schools.

PROJECT ACHIEVEMENTS

The project's successes are numerous. We now boast four different green travel initiatives that each week are used by 40,000 children and families at 200 primary schools in Kent, Medway and Bexley.

In the last academic year the impact of this initiative comprising walking, cycling, using scooters or Park and Stride (drive the first part of the journey and walk the remainder) was that 208,000 school run car journeys were removed from south-east roads.

A key part of the development has been the creation of a new initiative called Green Footsteps which is a low cost initiative that can be provided to a school within days of a school signing up and is able to be used anywhere in the country. This is being tested with south London schools prior to roll-out.

It operates hand-in-hand with a new cloud-based data capture area called the KM School Portal, that allows data to be input by schools and in return provides Ofsted self-evaluation evidence for head teachers. This combination of elements has enabled a big expansion of our work in Kent and a platform to expand.

We now have real momentum and after 15 years of operation, we are just starting to see the children of previous walk to school children helping to champion the green travel message.

WHAT DID THE PROJECT INVOLVE DOING?

The challenge we have faced as a charity is that our initiatives are quite resource heavy. Schools have a deadline to order the resources by and if they miss the deadline, they miss participation for a year unless we gamble and buy spares.

Creating a 'lite' version of the walk to school initiative called Green Footsteps has given us a lower cost and lower resourced scheme that a school can take up immediately should they miss the deadline for the main initiatives, or to use as a low cost option. Outside of Kent where postage costs might be an issue. Green Footsteps is the initiative of choice we offer to schools.

Alongside this was the issue of data capture. We needed a Cloud-based system that could link to our CRM database so our staff can instantly see headline figures for a school when they call in, while our partners such as councils can use the system to access detailed data on a class by class, school or district basis.

But we wanted the new KM School Portal to feel more like an information hub with documents and video posts available. Critically it needed to help sell the virtues of the initiatives, which it does with the My School's Ofsted Evidence button.

Clicking this generates cut and paste suggested Ofsted self-evaluation evidence that mail-merges in the headline stats for green journeys, cars removed from local roads and road safety skill test passes relevant to the school.

The aim is that even head teachers who are not passionate about the walk to school campaign, or perhaps those whose entire focus is ticking boxes for Ofsted, can find value from what our charity offers schools.

The next challenge has been testing this as a 'remote service'. It is one thing to pop down to the local school for a face-to-face meeting to sort an issue, but can we deliver an effective service at a distance. The test has been on four schools in south London and after almost a year all looks good so we are ready to take our message and services further afield.

Child's play: youngsters are urged to get their parents to start walking

WHO AND WHAT HAS BENEFITED?

The secret of our success is incremental change. If you say to a parent that they are never allowed to drive their child to school again they will laugh at you, of course. So our system creates an environment in the school community where child pester power can slowly nudge behaviour change in parents.

We do this through awarding our Green Class of the Week Trophy every week in every participating school. Children are desperate to win this and will pester mum or dad to walk 'just once a week' on our weekly survey day to increase the chances of their class winning it. Asking a parent to walk just once a week is our proven 'thin end of the wedge' technique to changing their transportation patterns.

Dad thinks he is not giving in much as he will still do the school run four days a week, while supporting his child with green once a week. But as the weather improves during the year and the walking habit gets entrenched, and children's pester power to walk kicks in, our families find additional walking days happen more and more regularly.

Now much of our activity is about empowering children to help us to deliver the green travel message to assist 'pester power'. Our new Green Champion initiative tasks School Councils with driving the message through simple child-led, whole school initiatives, such as newsletter creation and poster competitions. The best School Council ideas earn the children limited edition Green Champion T-shirts.

So many layers of additional service are now available to our schools compared to when you would have last looked at us. We really do feel like we have come of age and are ready for the big time!

KOREA AIRPORTS CORPORATION

LED-ILCMS DEVELOPMENT FOR CARBON REDUCTION

PROJECT AIM

Due to increasing greenhouse gases and global warming phenomenon, as well as abnormal weather, human security is being threatened. To reduce greenhouse gas emissions – the main culprit of global warming, the Korea Airports Corporation has carried out research and development projects for an eco-friendly green airport operation.

Among these efforts, an aviation equipment namely ILCMS (Individual Light Control Monitoring System) has been developed and applied to domestic airports. The company has successfully developed LED-ILCMS between 2011 and 2014, which is inter-operable with LED airfield lights, and applied this system in Gimpo International Airport.

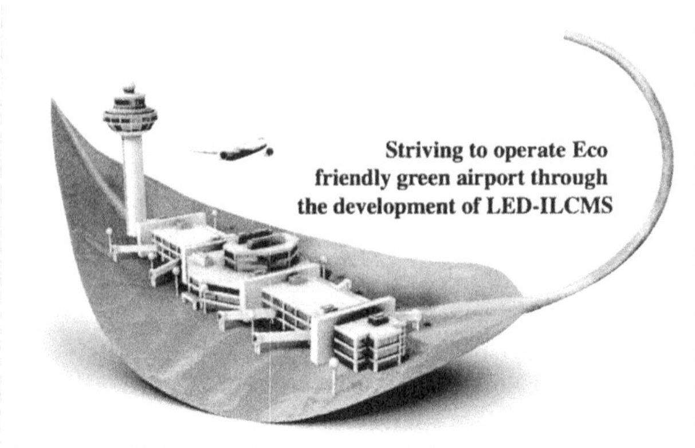

ILCMS is a system that can individually monitor and control airfield lights necessary for the take-off and landing of aircrafts based on advanced power line communication technology

Before the development and application of ILCMS, since all airfield lights had to be turned on, there was a serious waste of energy in operation. Also, as international standards require that the lighting rate be monitored and checked if there are at least two consecutive failures during low visibility, the project was started to improve the inefficiencies when checking the status of airfield lights using a vehicle during the flight.

PROJECT ACHIEVEMENTS

If ILCMS is applied at airports, energy consumption can be reduced as it can selectively light the airfield lights where the movement route is. Also, it can contribute significantly to protecting the global environment through valuable resource savings thanks to the extended life of airfield lights.

Above all, as for the aviation lightings for central line of runway or taxiway, thanks to the possibility to select lighting as per the direction of the aircraft, it can save as much as 50% power compared to the existing two-way lighting operation.

If the system is applied to the 7,300 of world's major airports, 630,720MWh of energy can be saved and airports can reduce annual greenhouse gas emissions up to 294,072tCO_2.

Saving energy as the system is changed from entire lighting operation to selective lighting for each direction (50% reduction) – 48W

Number of runway and taxiway centre-line lightings among 12 million of lightings in the world's major airports – 3.6 million (30% reduction)

Savings on taxiway centre line lights and airstrip lights × number of reduced lights × average number of operating hours = 48W × 3,600,000 lights × 10 hours × days = 630,720 MWh = 630,720MWh × 0.466249 = 294,072 tCO_2/year.

WHAT DID THE PROJECT INVOLVE DOING?

Korea Airports Corporation has been making various attempts to reduce greenhouse gas emission. As a result of these efforts, we has received a gold medal for the development of LED airfield lights in 2013 and a silver medal for the development of a power supply (AC-GPS) for aircraft ground in 2014 from the Green Organisation . Korea Airports Corporation has been continuously taking on challenges in the field of increasing energy efficiency, reducing greenhouse gas emissions at airports and trying to realise a friendly airport environment through research and development and investment.

WHO AND WHAT HAS BENEFITED?

ILCMS (Individual Light Control Monitoring System) can individually control and monitor airfield lights using power line communication technology applied to existing power cable but without any additional control wire or power line. As it can individually control and monitor the airfield lights using existing power cable without any additional controlling cable or monitoring cable, it is an environmentally friendly technology that can avoid additional cost caused by pipe and cable construction and resource destruction.

In addition, pilots can receive accurate and quick guidance for the take-off and landing since the airfield lights are only on at the movement route. With these benefits, it reduces not only the latency of the flight both on the ground and in the air, but also cuts the massive amount of greenhouse gas discharged from the aircraft.

The development of such a difficult technology requires the convergence of both professional expertise in aviation lighting and expertise in power line communication.

To realise its eco-friendly airport operation, Korea Airport Corporation has invested a lot of effort and money to develop LED-ILCMS, which is applicable to LED lights, since 2005 when ILCMS development project began.

The Korea Airport Corporation signed a joint development agreement regarding LED-ILCMS with power line communication specialists in 2006, and through continuous research and development, it successfully developed ILCMS in 2008. In 2014, it also succeeded in developing LED-ILCMS that is interoperable with LED lighting.

For development, they had conducted a load test using the test product during the years of technology development at the laboratory. Also for the research and development, they carried out overnight field test at the airport more than 100 times, and after 100 days of site operation test at Gimpo International Airport for system performance verification, they have successfully developed the product.

In addition, the technology was determined as a successful development from the technical evaluation agency, the Korea Industrial Technology Evaluation Agency, and was awarded a certificate from national test research institutes. Ten different patents of the results of the technology research and development have so far been registered.

For the commercialisation of the developed system, Korea Airports Corporation has contributed much to the development of technology and operations of environmentally friendly airport by investing 10 billion won and R & D personnel up to date.

In the world, only a few companies such as Honeywell have been able to develop such technology. In addition, we can say that it's more meaningful since Korea is the first successful case of such development among the developing countries.

WHAT ARE THE LONGER TERM BENEFITS?

Korea Airport Corporation has significantly contributed to greenhouse gas reduction and energy savings by installing ILCMS on the airfield lights in Jeju International Airport (2009). It also installed ungraded LED-ILCMS for a total of 5,652 airfield lights at Gimpo International Airport.

In addition, the application of ILCMS enables a scientific method in which aviation lightings can be remotely monitored, compared to the existing method in which visual inspection by car around the landing strip is required. As a result, it can greatly contribute to the reduction of greenhouse gas emissions and air pollution prevention.

In October 2014, the company operated a booth at an inter-airport international exhibition held in China and tried to apply the scheme to other overseas international airports. It is expected to continue to expand and apply to the world's major airports.

The project started in 2014 and is expected to progress with 19.4 billion won of the national budget by 2018. A-SMGCS, based on LED-ILCMS, is able to monitor any moving object within the airport by location information linked with flight information, weather information, and ground surveillance radar from aircrafts near the airport.

As the system can automatically guide pilot to the best route, Korea Airport Corporation is getting closer to the realisation of operating green airports.

These kinds of research development efforts and investment from Korea Airports Corporation will contribute to the friendly airport operation around the world, and further development and expansion to the world can be expected.

KOREA ENVIRONMENTAL INDUSTRY AND TECHNOLOGY INSTITUTE (KEITI)

GREEN CREDIT CARD

The Green Credit Card is an initiative led by the Korean Ministry of Environment (MoE) and Korea Environmental Industry and Technology Institute (KEITI) in partnership with the private sector to mitigate CO_2 emission through demand-side management.

Economic rewards are provided to credit card users for purchasing eco-friendly products, using public transport and saving utility rates including electricity, water, and gas at home.

The economic rewards are converted into cash points called eco-money points and accumulated on the credit card system, later to be redeemable for cash or used for various purposes. The accumulated points can be donated to environmental funds aimed at mitigating climate change.

The Green Credit Card aims to contribute to reducing CO_2 emissions by offering tangible economic incentives for low-carbon lifestyles and consumption patterns. Furthermore, it is intended to nurture markets for greener products, thus driving eco-innovation to underpin a green economy.

PROJECT AIM

The Republic of Korea is the 12th largest emitter of greenhouse gas in the world, and the largest portion comes from electricity and heating (37%), followed by manufacturing and construction (14.5%), and finally the transport sector (14.2%).

The Korean government recently renewed its commitment to curb its GHG emissions by 37% relative to BAU by 2030, and various policy instruments have been undertaken to decrease its GHG emissions and pursue sustainable development.

The Green Credit Card is an initiative led by the Korean Ministry of Environment (MoE) and Korea Environmental Industry and Technology Institute (KEITI) in partnership with the private sector to mitigate CO_2 emission through demand-side management.

Economic rewards are provided to credit card users for purchasing eco-friendly products, using public transport; and saving utility rates including electricity, water, and gas at home.

The economic rewards are converted into cash points called eco-money points and accumulated on the credit card system, later to be redeemable for cash or used for various purposes.

The accumulated points can be donated to environmental funds aimed at mitigating climate change. The Green Credit Card aims to contribute to reducing CO_2 emissions by offering tangible economic incentives for low-carbon lifestyles and consumption patterns. Furthermore, it is intended to nurture markets for greener products, thus driving eco-innovation to underpin a green economy.

PROJECT ACHIEVEMENTS

The Green Credit Card was designed in partnership with various public and private partners. Seventeen financial institutions, 26 retailers, 160 manufacturers, and 17 municipalities are affiliated with its operation.

As a result, economic rewards equivalent to 5.2 million USD have been offered for a wide range of eco-friendly consumption and behaviors including the purchase of 1,431 eco-friendly products sold at 34,825 retail stores; the use of public transportation including subway, intra-city bus, inter-city bus, and bullet trains nationwide; and saving utilities at home including electricity, water and gas.

The Green Credit Card is one of the best selling credit cards in Korea, attracting about 9.4 million card holders. As of April 2015, the accumulated number of card holders exceeded ten million, which is equivalent to 37 per cent of the economically active population of Korea. The sales of eco-friendly products paid with the Green Credit Card increased 1.6 times from 2013 to 2014, and the total sales of eco-friendly products amount to 4.4 million USD.

Public transportation expenses paid with the Green Credit Card totalled 271 million USD from 2011 to 2014. On average, the Green Credit Card holders take public transportation using the Green Credit Card about 3.3 times per month, which outperformed other credit cards in terms of frequency of monthly use of public transportation.

Economic rewards totalling 2.3 million USD have been provided to 140,681 households for saving utilities at home. And the number of local public facilities (e.g. national parks, museums, etc) that offer discounts on entry fees to the card users increased from 178 in 2011 to 762 facilities in 2014.

WHAT DID IT COST?

Initially, 1.6 million USD was invested by the operating agency (BCcard) in setting up a Green Point of Sale (POS) at more than 34,000 retail stores of 20 retailers, which account for 80% of the entire retail market of Korea. Yearly, about 2 million USD is spent to operate the system.

WHAT DID THE PROJECT INVOLVE DOING AND WHAT DID YOU DO IT?

The MoE and KEITI are responsible for the establishment and operation of the overall system. Firstly, credit card reward points called Eco-Money Points, are given for the purchase of eco-friendly products with the budget of manufacturers and KEITI. When an eco-friendly product is purchased by the Green Credit Card at a store equipped with green point-of-sales (POS) system, 3 to 24 percent of the product price is offered by KEITI as reward points.

Green POS system

Green point of sale (POS) is an integral part of the Green Credit Card system. It enables stores to automatically recognise green products via a barcode scanner, and award reward points to consumers when green products are purchased with the Green Credit Card. Green POS builds an online database on green products, managed by KEITI, which provides information on green products.

Second, reward points are provided by the operator, BCcard, for use on public transportation. Up to 20 per cent of transportation expenses are accumulated as reward points for the use of the bus or subway network. Five per cent of transportation expenses are accumulated for the use of bullet train or express bus.

Third, points are offered with funding provided by the MoE and municipalities for saving utilities such as electricity, water or gas at home. The reward points are calculated based on the date of the reduction of electricity, water, and gas (LNG) during the period of the past six months, compared to the average uses of the respective utilities for the previous two years.

Fourth, the cardholders are allowed to use public facilities such as national parks and museums free of charge or at a discount price.

WHAT IS THE SUSTAINABLE IMPACT?

The sustainability impacts generated from using the Green Credit Card are equivalent to an economic savings of 157 million USD. About 531,000 tons of CO_2 equivalent emission were reduced from the onset (July 2011) to Dec 2014. The benefits can be broken down into the following:

- A total of 47 million litres of gasoline hae been saved by the Green Credit Card users through switching their mode of transport from private vehicles to public transport
- About 223 million kW of electricity, seven million tons of water, and 11,000 tons of gas were saved by the Green Credit Card users at home
- In total, economic savings of 141 million USD resulted from the reduced use of gasoline from the use of public transport as well as savings of electricity, water, and gas at home
- About 775,000 USD of environmental costs was saved from the prevention of gasoline production and combustion required for the use of private vehicles
- About 14 million USD of environmental costs was avoided from the prevention of the production of electricity, water and LNG and the combustion of LNG
- About 20,000 tons of CO_2 equivalent emission reduction has been made from the purchase of eco-

friendly products using the Green Credit Card, which is equivalent to an economic savings of 123,000 USD.

WHO AND WHAT HAS BENEFITED?

Consumers benefited from the convenient and eco-friendly credit card reward system. When payment is made with the Green Credit Card, about 10.3 percent of the total payment amount is accumulated as Eco-Money points. Up to 10 USD of discounts on public transport expenses can be provided to the card users per month. In addition, economic rewards gained from saving utilities at home are automatically accumulated on the credit card reward system for the users' convenience.

Manufacturers, retailers, and credit card companies benefited from the increased sales of eco-friendly products by 1.6 times last year.

Local communities benefited from the increased visits of the card holders to local facilities from 744 in 2011 to 77,291 in 2015. The general public benefited from cost savings in resources including energy, water, and gas and the improved environment by avoiding the emission of environmental pollutants.

WAS THERE ANYTHING INNOVATIVE ABOUT THE PROJECT?

As an economically viable and and eco-friendly incentive scheme, the Green Credit Card makes economic sense by converting the profits generated from the sales of eco-products into eco-money points. It is financially viable and does not require substantial funding from the government.

Public/private partnership business model contributes to sustainable development. A variety of stakeholders including consumers, retailers, manufacturers, and credit card companies participate in and benefit from the system.

The use of a mobile application for customer relations means that any information on eco-friendly reward points can be accessed by the respective cardholders through the mobile application in real time. In addition, the card users can check on useful information including a list of the eco-friendly products and retail stores with the Green Credit Card, and event promotion activities.

CAN OTHER ORGANISATIONS BENEFIT FROM IMPLEMENTING YOUR METHODS?

The Green Credit Card has been presented as an example of good practice to spur eco-friendly consumption and lifestyles at numerous international meetings, publications, and news articles. The Green Credit Card has a potential to be replicated on various scales, depending on the respective context and policy needs. Eco-friendly rewards can be provided with or without the linkage of credit card for a wide range of eco-friendly lifestyles and behaviors at local, national, or global levels.

There are growing demands from abroad including China, Thailand, and Helsinki in Finland, to benchmark the Green Credit Card. A number of training sessions were held by KEITI to transfer the knowledge and experiences to countries that wish to introduce the similar system. A Memorandum of Understanding was entered into between KEITI and the National Strategic Union on Haze Prevention & Control (NSUH) of China in order to enhance cooperation for introducing the Green Credit Card in China in April 2015.

WHAT LESSONS HAVE YOU LEARNED?

The Green Credit Card has been effective in closing the information gap and tackling price barriers faced with eco-friendly consumption. Lack of information as well as relatively high prices for eco-friendly products was found to be major barriers hampering the consumption of those products.

The price competitiveness of eco-friendly products increased as a result of the provision of economic rewards up to 24 percent of a product price. Consumers could identify eco-friendly products more easily through the indication of eco-friendly products at stores by partner retailers, and the mobile application providing a list of the affiliated eco-friendly products.

Eco-friendly economic rewards should be provided in a user friendly way. The credit card is the most commonly used medium of payment for public transport as well as purchasing products in Korea. Therefore, linking an eco-friendly reward system with a credit card platform was critical in designing a convenient eco-friendly incentive scheme for consumers.

Engaging and coordinating various stakeholders, including retailers, municipalities, and manufacturers, etc are the key success factors in the operating the system.

ARE YOU PLANNING ANY FURTHER DEVELOPMENT?

Expanding the service network in order to provide a broader range of eco-friendly economic rewards: The MoE and KEITI plan to engage various types of retailers and increase the scope of affiliated eco-friendly products and lifestyles, including paperless receipts, car sharing, and reducing food leftovers at restaurants.

Big data analytics will be conducted in order to provide customized eco-friendly rewards and information that further promote eco-friendly consumption and lifestyles.

Coordinating with other governments' vouchers in order to enhance users' convenience: The MoE is coordinating with the relevant ministries including the Ministry of National Defence and the National Health Insurance Service in order to come up with an integrated green credit card that offers voucher services provided to soldiers, mothers with children, etc.

By doing so, consumers do not need to register for different credit cards services managed by the different governments.

Developing an online and/or mobile version of the Green Credit Card: Online green POS is being developed in order to offer eco-money points for online shopping of eco-friendly products. In addition, efforts are being made in order to link the Green Credit Card with mobile payment.

Building a global partnership for the Green Credit Card: Korea wishes to set up a global network of the Green Credit Card, thereby enabling the accumulation and the use of rewards for the purchase of eco-friendly products and services across the globe.

A certain portion of the payment made with the Green Credit Card will be channeled into a fund to contribute to meeting sustainable development goals as well as tackling climate change.

COUNCIL OF MINISTERS, KUWAIT

JLEEB CONSTRUCTED WETLAND, KUWAIT

PROJECT AIM

Our emphasis was on an approach to find environmental and economical venture by any environmental method for sustainable use of earth's valuable resource – water, which is the most constrained natural resource especially in our desert environment in Kuwait.

Our aim was to build an optimal system for the treatment of waste water onsite and use its yields as a sustainable resource for irrigation needs, with reduced involvement of mechanical process for water reclamation.

Desert bloom: ariel view of the wetland in dry Kuwait

We were to undertake a sustainable development project to conserve and reuse waste water at the source, which was being discharged into the municipal main sewage line. (approx 90 m3 daily).

We also opted for finding an alternate source for the portable water needed for an environmental improvement project to water a 3,500 m2 plot of land planned for a lawn to serve as a recreational area for residents of the vicinity.

A technology offering us range of alternative solutions to supports our plans for the management of water within sustainable environmental, economic, and social conditions was introduced by a German company to construct a wetland.

PROJECT ACHIEVEMENTS

The construction and operation of the constructed wetland project in a desert environment indicated many challenges in the development of a sustainable process of waste water reclamation and reuse.

A successful construction and operation of the onsite constructed wetland for the past seven years has fetched multi-dimensional benefits to the environment, nature, economy, recourses, and the community. Important project achievements recorded over the year were;

- Decentralise the discharge of municipal waste water from residential buildings to reduce the load of waste water on the conventional treatment plant (daily average – 90 m3)
- Introduced the reuse of treated water, high in nutrients from our system for landscape irrigation (daily average – 77 m3)
- Created a lush green lawn area for recreational purposes with negligible dependence on portable water
- Saved on the logistical cost for transportation of fresh water from water depot intended to irrigate the lawn area
- Reduced carbon foot prints for energy used in conventional treatment plants, desalination plants, fossil fuel usage and oxygen production by plantation
- It was possible to construct the wetland with mostly local materials and equipment for ease of further projects
- Very low maintenance and repair cost occurred for the smooth operation of the constructed wetland plant

- The wetland as a biological system fixes carbon in the plant structures through photosynthesis
- Major reduction in water pollutants – chemical, biologic, and heavy metals contents.

WHAT DID THE PROJECT INVOLVE DOING?

"Sustainable development is development that meets the needs of the present without compromising the ability of future generations to meet their own needs." – Brundtland Commission (1983)

Water is one of the world's most valuable resources but it is under constant threat due to climate change and resulting in drought and rapid population growth.

Kuwait is located in one of the driest region within the Arab Gulf states.It has an arid climate with conditions being hot and dry with a total mean rainfall rate of about 110 mm a year.

Kuwait's water consumption can be compared to any metropolitan city located next to the sea in a desert location. The primary source of fresh water is through its desalination plants producing portable water from sea water.

High water consumption results in large volumes of waste water from the urban lifestyle having to be treated in conventional treatment plants, all leading to economic, environmental and social costs with increased carbon foot prints at every stage of process.

WHY DID YOU DO IT?

Careful and effective water management needed consideration to enable water conservation and reclamation under sustainable economic, social, environmental conditions.

The Water Reuse Association defines reused, recycled, or reclaimed water as "water that is used more than one time before it passes back into the natural water cycle". Thus, the environment benefits extensively from the recycling and reuse of treated industrial and municipal water for beneficial purposes such as agricultural and landscape irrigation

Sustainability challenged us to reflect on waste water treatment differently. Instead of focusing on end-of-pipe-treatment for emission prevention, attention shifted towards optimal resource utilisation, favouring the development of decentralised systems.

An extensive study and review of green technologies gave us an overview into selecting one such sustainability methods of water by means of a constructed wetland. This methodology permitted us to build an optimization system to treat wastewater onsite and use its yields as a sustainable water source.

Our aim was to focus on the municipal wastewater being discharged into the sewage mainline from the staff accommodation for 500 residents, which is normally treated at conventional treatment plant.

To treat the wastewater resulting from this neighbourhood, it had been planned and designed to implement the concept of constructed wetland according to the local climate, conditions and material availability.

Emphasis was given to the correct management of water, both in terms of quantity and quality, with minimal dependence on mechanical processes of treatment of wastewater.

The attractive method of water reclamation by means of reed beds in a constructed wetland was to be benefited by the reclamation on average of 90 m3 (20,000 gallons) of waste water every day at a low cost of construction and operation. The concept of building a constructed wetland was to process the volume of waste water being discharged by the residents on the allotting area of 1,800 m2 for filtration beds.

Alongside the wetland, we had the opportunity to develop a proposed recreational area of 3500 m2 for the residents as a recreational area and to utilise the yields from the waste water treatment onsite for landscape irrigation.

Our emphasis was on an approach to find an environmental and economical venture with a sustainable method for the use of earth's most valuable resource. The aim was to build an optimal system to treat the wastewater onsite and use its yields as a sustainable resource for irrigation purposes, with reduced dependence on mechanical process both for waste water and fresh water.

Our second aim was to find an suitable alternate source to our requirement of portable water needed to irrigate a plot of land to be developed as a lawn to serve as a recreational area planned for the vicinity. Thus, a technology offering a wide range of alternative solutions was introduced by a German company to construct a wetland.

WHAT DID IT COST?

The constructed wetland project in Kuwait was intended to experiment the adaptability and the applicability within the local environment. Our objective was accomplished by the design, construction, operation, maintenance and performance with the total cost of construction of the wetland at 71,820 Kuwaiti Dinars (£159,500).

The other costs involved for the operation and maintenance of the wetland were low and the cost for both the treatment plant and the landscape area averaged at 5,500 Kuwaiti Dinars (£12,250.00) per year.

The project had been funded by the Environmental Tri-Committee under the Council of Ministers for the Kuwait Government's initiative to invest in environmental sustainability pilot projects to encourage new technologies for environmental improvement in Kuwait.

WHO AND WHAT HAS BENEFITED?

This project achieved many targets set for environmental sustainability and development for various aspects for the environment in Kuwait.

The project was able to reclaim approximately 163,447 m3 of municipal wastewater in its first five years of operation. With the operation of the constructed wetland it was possible to reuse 140,525 m3 treated water in the same period.

An alternate to the portable water requirements for the lawn area was available so it was possible to save 140,000 m3 of valuable fresh water intended for landscape irrigation. A well-maintained 3500 m2 area for a lawn with lush green grass and plants was possible because of the this method of treatment and at this low cost.

The setup of the wetland operation raised some questions about the system's suitability to the harsh hot climatic conditions and its stability and constant pollutant removal efficiency, a sustainable source of treated water for reuse for irrigation and the limited logistic costs of water treatment and irrigation. The treatment process has operated efficiently for the past seven year with remarkable benefits and we hope it to continue for many years to come without failure or major repairs.

The construction of wetland benefited the 500 residents of the neighbourhood with a lush green area to be used for their recreational needs. The wetland has attracted a lot of birds and animals looking for respite in the hot and dry weather.

Other benefits are:

- The recovery of carbon dioxide to oxygen from the reed plants, trees and grass spread over 3500 m2 of the lawn area created by the reuse of treated water
- Reduction in load on the water resources both in terms of wastewater treatment and fresh water production, by the repeated use of water for different proposes
- Overall environmental benefits incurred for the improvement and development for sustainable measures for the local community.

WHAT ARE THE LONG TERM BENEFITS?

The project of constructed wetland adhered to providing a sustainable source of water from waste water reclamation with a nominal investment compared to its yields and lifespan.

With minimal dependence on portable water, a valuable green recreational area has been developed and can be used by many residents for the natural life of wetland operation.

This is a sustainable method for environmental improvement for use and reuse of water before it is introduced into its natural water cycle. Even though wetland both natural and constructed are a regularly occurrence in Europe, our wetland was unique for being the first of its kind in Kuwait and this desert region.

Our project has been at the centre of attention since its operation as it was under observation for its level of performance in an arid region, which generally lacks natural or constructed wetlands. The construction concentrated on naturally and locally available materials with little dependence on imported materials.

The environmental aspect associated with this method of waste water treatment have set important indicators on means to protect sources of drinking water in this region.

CAN OTHER ORGANISATIONS BENEFIT?

The process of the wetland set an example of waste water treatment by natural means and has attracted interest from organisations and communities to implement this technology at various locations with multiple uses.

We have worked on different fronts to explain the results that had been achieved with the implementation of this method for water treatment for onsite operations, with low construction, maintenance and operation costs, especially for remote areas of Kuwait.

Being a government initiated project the wetland has been visited by government departments, environment organisations, research centers, colleges, schools, etc. We were able to gain extensive knowledge in this technique of water reclamation and reuse especially in a desert location, where water seems to be the most valued natural resource.

We have been working with different public and private sector organizations to utilise this system in remote area of Kuwait where either sewage mains are not available nor connected to the national grid.

We are in consultation with the oil industry to introduce this technology, where water is being used in great volumes during the drilling processes of oil.

We have proposed to the Kuwait border and security department to build wetlands for the border outposts, where access is always difficult.

We are constructing a wetland project for the Institution for Environmental Studies so it would be possible to educate students with this technology.

Upon successful completion of the operation for continues five year, an endorsement was acknowledged by a Project Evaluation Report from Hochschule Ostwestfalen-Lippe, University of Applied Sciences – Department of Civil Engineering, Institute for Sanitary Engineering, Detmold, Germany.

LAND SECURITIES

GUNWHARF QUAYS WASTE MANAGEMENT

PROJECT AIM

To be a leading example of how a large waste producer can have a substantial positive impact on waste management performance, when each waste stream is thoroughly reviewed, policies are adopted and systems are installed, which focus on continuous improvement and zero tolerance to environmental impact. All these initiatives should provide a substantial impact on the sustainability of Gunwharf Quays.

Our objectives for 2014 to 2015 were to increase the volumes recycled and reused and to lower the amount of waste sent for energy recovery. With a target for the year of 30% to the Energy Recovery Facility (ERF), we finished the year at 31%.

Gunwharf Quays sends no waste to landfill sites. We aim to be an example to other businesses by ensuring our reporting is as close to 100% accurate as possible and also transparent, based on individual on-site weighing and comparing against waste transfer and weighbridge notes, compliance documentation and with reuse materials and items being individually signed for. The aim was to remove the use of average weights for the majority of our waste streams.

PROJECT ACHIEVEMENTS

In 2014/2015 Gunwharf Quays produced 3,203 tonnes of total waste, with 69% (inclusive of 3% reuse) or 2,118 tonnes of that figure being recycled .

In comparison with 2013/14, 52% of total waste was being sent to the ERF, giving us a total waste recycled figure of 44% and 4% re-use of 2,858 tonnes total waste. By the end of 2014/15 we sent 481 tonnes less waste to the ERF than in 2013/14 at a time when total waste from the centre had been increasing.

We presented to our 134 Brand Partners (BP) over four days, introduced Environmental Waste Forums, employed a Waste Champion and conducted a full review of all our waste streams, to confirm what was being disposed of, by which BP, and what we could do to maximise re-use and recycle materials.

Our objective for 2015/2016 is to increase reuse and recycling to a combined 75%, thus decreasing our waste to ERF to 25%. We are on target, with our May performance increasing to 75% reuse/recycling and 25% to ERF. We continue to review, amend and create a tailored approach to our waste management on site which provided substantial savings of £60,450.14 in 2015/2015.

WHAT DID THE PROJECT INVOLVE DOING AND WHY DID YOU DO IT?

In 2013 Land Securities reviewed the overall waste performance of the business and while already having objectives and results following zero waste to landfill, we were confident that further improvements could be made.

A directive was issued to all centres to review how they managed their waste and set new objectives as a company, to increase the amounts reused and recycled.

Gunwharf Quays is a premium retail centre and a flagship centre in regards to the Land Securities retail portfolio. As the flagship centre, we felt we could be a great example to our business and to the industry, as to just how far you can push the performance of waste management and minimise its impact on the environment.

We conducted a full review of all of our waste streams in detail to confirm what was being disposed of and by which BP; and what could be done to eradicate wasteful management and ensure we maximise reuse and recycle materials.

In order to identify the contamination, our Soft Service provider broke open black sacks on a daily basis and logged the offending BP and would then discuss with them direct the required areas of improvement.

To make this task less time intensive we installed a policy of clear bag use only and would not service any areas where black bags were used for waste disposal. We put together the following waste streams overview:

- Existing: general waste, cardboard, LDPE (polyethylene), glass, metals, pallets, coat hangers.

We confirmed we could handle the following additional waste streams and over the course of2014/15 have improved our systems for handling, storage and changed suppliers to ensure efficiency, maximise resource retention and obtain transparent reporting of data:

- Reuse materials: WEEE, food waste, DMR, plastics PET/LDPE/HDPE/PP/PS, textiles and timber.

Our most notable successes this year are as follows:

- Reuse items/materials

To date we have reused 80.1 tonnes, a combination of WEEE, textiles, coat hangers, mannequins, shelving, furniture, suitcases and pallets. WEEE Is now reused or recycled through a charity (Computeraid) and sent to developing nations for use in schools and hospitals.

- Food waste

We have 39 producers of food waste. In 2014/15, 100% of our food waste was sent to the ERF. Our food waste management now represents 20% of total waste recycled

- DMR

(Coloured paper, tins, cans, plastics) are collected in DMR bins and tipped into a 35-yard enclosed compactor. In particular to our 95 BPs on the centre, we are now utilising the new Waste Regulations January 2015, to assist us in enforcing a zero contamination policy.

- Plastics PET/LDPE/HDPE/PP/PS are separated from general waste, pallet wrapped and collected by a waste broker
- Textiles

In agreement with our BPs, tags are removed to make them unsellable. We believe we can greatly improve our volumes which are recycled or directly reused by local charity partners.

- Timber

Reuse is dependant on timber quality, otherwise it is chipped and sent to be used in flat pack furniture.

- Metals

These are stored in a 35 yard bulker in our service area. The metals are melted down and reused in manufacturing. We now receive a rebate on all metals based on letsrecycle.com, which is reinvested in to innovations.

- LDPE & cardboard

Bale sizes of 450–650 kilograms, dependant on density and the machines configuration. These are stored in our service yard for collection by our waste broker. We receive a rebate of £200 & £60 respectively.

Corporate social responsibility

Following this review, we brought on a number of new waste providers and charity partners which would both support us and allow us to also support our corporate social responsibility aims. To date we have supplied all of our reuse items (excluding pallets) and some recyclables (ICT components) to charity partners at The Rowans Hospice, Mencap, The Beneficial Foundation and Computeraid.

WHAT WAS INNOVATIVE ABOUT THE PROJECT?

All savings are reinvested in environmental performance and innovation. We have invested in a waste trailer to allow us to be more efficient in the handling of our fleet of bins (400) across our 33 acre centre. This has provided a saving to labour hours as well as saving on bin consumables, hitches, wheels etc.

We used to utilise a Kawasaki 4010 diesel for towing 2 x 1100 litre bins across the centre. For the waste trailer we required a vehicle with a larger towing capacity. We therefore invested in a Polaris Ranger utility vehicle, which is able to tow up to one tonne at a time.

We now use the Kawasaki for towing a bin cleaning trailer with water bowser, which was purchased from savings made. To better manage our waste contractors/couriers we purchased 4 x LCD weighing pallet trucks which allow us to weigh bins/recyclables ourselves. This allows us to monitor the figures we receive (to +/- 1kg accuracy) from our contractors by spot checking, to ensure we have accurate and transparent reporting of our performance.

Through the review of our waste suppliers, we have requested the use of live weighing systems on collection vehicles, which will soon be rolled out on various streams.

By utilising a customer zone we can compare collection figures with our data by 5pm that day. This ensures that we have a further representation for transparent reporting and greatly supports our aims to make improvements.

IS THE INITIATIVE COST EFFECTIVE?

In comparing the cost effectiveness of recycling versus our historical ERF arrangements, it can be misunderstood that having a larger number of waste contractors/partners may incur a greater cost rather than having one supplier with economies of scale. Or it may be perceived that a separate arrangement for different waste streams would require more consistent management time.

However, by installing the correct policies, structure, processes, procedures and waste contractors, we have shown a substantial amount can be saved on our waste management budgets, which we have chosen to reinvest to drive forward our environmental performance.

MARINE RESEARCH FOUNDATION

PROMOTING AN ECOSYSTEM-BASED APPROACH TO FISHERIES VIA TURTLE EXCLUDER DEVICES IN MALAYSIA

PROJECT AIM

As yet there is no ecosystem-based fishery management in Malaysia and there is scant involvement of local fishing communities in fishery management decision-making and policy. Because of this, bycatch is a serious concern – particularly for large, endangered sea turtles, which are drowned in shrimp fisheries at a rate of some 2,000 to 3,000 per year.

Safer swimming: TED is key to giving turtles a better chance of survival

Our goal is to create an effective partnership with the Malaysian government and with fishing communities across the country, and help them adopt Turtle Excluder Devices in shrimp trawls, which will save sea turtles while still allowing fishing to continue.

A TED is basically a metal grill that allows the catch to pass through a net flap, but ensures that the sea turtle is unable to get through. TEDs also save on fuel (increasing benefits to fishers) and reduce damage to the catch (again increasing benefits to fishers). More importantly, they save sea turtles, which are classified as endangered in Malaysia.

Our goal is not to shut down fisheries but rather to make them more sustainable and more relevant in the face of ecosystem approaches to fisheries management (EAFM). The reduction in bycatch of endangered sea turtles, themselves key components of marine eco-systems, will lead to incremental benefits to the marine environment, while continuing to sustain fishing communities.

PROJECT ACHIEVEMENTS

This project started in 2007 as a small NGO-led initiative. Today, it is a growing and embedded national government effort with the full buy-in of the Malaysian Department of Fisheries (DOFM).

It is a wonderful account of how a small, determined and meaningful NGOs effort to save sea turtles was translated into real on-the-ground action, with a positive impact on sea turtles, and working hand in hand with fishing communities, not against them.

Through this project we developed a long-term national bycatch reduction programme in partnership with DOFM, which has resulted in an improved conservation status for sea turtles in Malaysia while ensuring continued fishery benefits. The key outcomes of our work include:

- Provision of critical knowledge needed for decision-making and effective conservation of sea turtles. We have helped DOFM understand the effect and magnitude of shrimp trawler bycatch on sea turtles
- Incorporation of sea turtle and ecosystem conservation priorities and measures into relevant policy - in 2017 TEDs will finally become a legal requirement in Malaysia
- Enhancement of local fisher community-based stewardship of sea turtles. This is where the true value lies as fishermen now use TEDs and turtles are saved each and every day.

WHAT DID THE PROJECT INVOLVE DOING?

Sea turtles in Malaysia are iconic as they are across the world's oceans. They have the power to melt young hearts, to intrigue scientists and to mesmerise and impress the public at large

But they can hardly be considered 'safe', as a suite of pressures, both foreign and domestic, threaten their very existence. Eggs are poached on remote islands. Large and small turtles are poached for meat and as curios. Nesting sites are lost to coastal development.

But the greatest threat to sea turtles is accidental capture in commercial and artisanal fisheries. As shrimp trawl nets roll along the seabed, they indiscriminately catch and drown numerous sea turtles – as many as 3,000 each year in Malaysia alone. But practical and inexpensive solutions exist in the form of Turtle Excluder Devices (TEDs), which are fixed within trawl nets and allow a fisherman's catch to be retained while turtles are excluded.

Although TEDs have many advantages, fishermen are wary of using them because of the large exit 'trapdoor' in the net that allows turtles and debris to escape. They are concerned that their target catch will be lost through this opening, even though self-closing netting flaps cover it.

They even blame TEDs for catch losses when these are unrelated. As I stated in a grant application once: "When fishermen have a good day, it's a good day. But when they have a bad day, it's the TED's fault!"

Our project has involved years of spending an incredible amount of time on fishing boats, working with fishermen and introducing Turtle Excluder Devices one boat at a time.

We trialled TEDs across Malaysia, and became adept 'fishermen' ourselves in the process. We spent time at sea, using TEDs, discussing their function and benefits, solving problems and fixing nets alongside the crew. We ate and slept and laughed and cried with them – and we were taken in.

We created a documentary video so our new family could understand TEDs better. We went the distance – figuratively! We took six key fishermen on a study tour to the US, hosted by the National Marine Fisheries Service.

We later took government officials and the Director himself on study tours to learn about TEDs. We put GoPro cameras on nets to prove to fishermen that the flaps were indeed closed and that turtles were being saved every day.

We convened national and state-level workshops to spread the word far and wide, and we helped fishermen come up with a new TED design that would be more acceptable to them – incorporating a 'smiley face' opening that allowed the capture of small stingrays – a key economic staple. More importantly though, we got the Malaysian government on board to drive the programme, because without legal backing TEDs were never going to make it to the 'big leagues'.

WHO AND WHAT HAS BENEFITED?

Over the last decade we have raised around half-a-million American dollars to fund this work, supported by Malaysia's GEF Small Grants Programme, the US NOAA National Marine Fisheries Service, the Save our Seas Foundation and Conservation International.

The ultimate beneficiaries are sea turtles, but also the fishermen. They get to save endangered species at no net loss to themselves. As the government embarks on development programmes, we did not want to be imposing 'western' conservation values on fishermen without them seeing the benefits this brought to them.

We are happy to say these things have progresses hand in hand rather admirably. We now have more sustainable fisheries and are saving endangered species daily. It's a true win-win situation.

And more so today as the Malaysian government has sought our help to address other bycatch issues and solutions, and introduce these to fishing communities in a similar manner. We recently had a GEF medium-size grant approved for $1,250,000 to support wider bycatch mitigation and sustainable fisheries issues. In the coming years we will be saving marine mammals and sea turtles and a suite of critical species, but always with the sustainability of the fisheries in mind.

TEDs are not new – they were invented by US fishermen some 40 years ago – but what has made our work unique has been the partnership approach, and the patience.

We knew things would not change overnight. Indeed, we knew a one to two year investment was a waste of money if we did not go the distance. Our strengths, as a foreign NGO, came in speaking the local language, being willing to put in the time and effort and discomfort and time away form home – but great fresh seafood dinners each night! – and also in being able to bring on board a suite of donors so that no one single body absorbed the entire costs of implementation.

As one donor moved on, we turned to another. In this manner we managed to always have support from year to year, and this mattered. We were able to be a presence in the fishing community on a continuous basis.

WHAT WAS INNOVATIVE ABOUT THE PROJECT?

The novel aspect of our work has been our ability to work closely with the Malaysian government in a climate where foreign NGOs can often be seen as interfering bodies rather than collaborators.

Our ability to highlight the benefits to all parties involved – government programmes included – was a mainstay of the TED programme. Today, our work helps the Malaysian government address its commitments under the CBD, FAO Code of Conduct on Responsible Fisheries, CMS's IOSEA Sea Turtle MoU, the Coral Triangle Initiative, the Malaysia Sea Turtle Plan of Action, and more.

For this project among others, MRF's Executive Director, Dr Nicolas Pilcher, was recognised as a Conservation Hero in 2011. Our TED-related work has featured in international press and in the local media, on television and in newspapers, and in books and articles.

MARMAX

SPECIAL EDUCATION NEEDS RANGE

PROJECT AIM

Our key outcome and objective was to provide award winning furniture and play equipment that was accessible to all, encouraging inclusive play, working in partnership with SEN Schools and seeking expert advice taking into consideration not just fitness and mobility but also behavioural and learning difficulties as well as emotional health and well being.

Table talk: youngsters take full advantage of the award winning furniture

As well as creating a range for those with special education needs, we also had to ensure that the range met with the requirements of the Sustainable School Strategy and Eco Schools programme. The range would have to:

- Capture their imaginations in outdoor play, and help them learn about sustainable living (Sustainable Schools)
- To become models of waste minimisation and sustainable procurement (Sustainable Schools)
- To create a richer teaching environment and enable pupils to identify links between topics (Eco-Schools England)
- To encourage pupils to value the school grounds as a place to play, explore and make a connection with the natural world (Eco Schools England.

PROJECT ACHIEVEMENTS

By opting to use recycled plastic, this ensured that the new Special Education Needs range would comply with the government's sustainable school strategy and the Eco Schools England programme of which most schools in England are registered. The Special Education Needs range encourages active, imaginative play, personal, social and emotional development.

The design of each product has been to include special needs accessibility for all regardless of ability so, where required, ramps have been incorporated to provide access for all and a range of easy accessible seating has been created, thus ensuring the range meets its purpose.

The range is practical in design and because it is made from 100% UK recycled plastic, it is sustainable, robust, durable, and easy to keep clean with no maintenance required. Each piece of play equipment also includes a unique anti-slip surface ensuring the safety of everyone that uses it.

The full Special Needs range is made from 100% HDPE UK post consumer waste, which is more commonly known as milk cartons and the full Special Needs range saves 57,914 two pint milk cartons from going to landfill. Each product in the range has been specifically designed to encourage inclusive play amongst children no matter what their ability and meets with the following government initiative.

WHAT DID THE PROJECT INVOLVE?

The project involved working alongside a local school to design and create a range of Special Educational Needs playground equipment and seating suitable for pupils with a range of disabilities.
After months of market research with architects, schools and with members of the education building forum, it became clear to the sales and marketing manager that there was a gap in the market and a need for Special Educational Needs products, and not just for schools but also for playgrounds.

Accessible play: Marmax has filled a gap in the market

The sales and marketing manager also discovered an initiative where the government were encouraging inclusive play. This confirmed there was a need for this range.

As well as creating a range for those with Special Education Needs, we also had to ensure that the range met with the requirements of the Sustainable School Strategy and Eco Schools programme.

By opting to use recycled plastic, this ensured that the new Special Education Needs range would comply with the government's sustainable school strategy and the Eco Schools England programme of which most schools in England are registered.

WHAT DID IT COST?

It is difficult to put a final cost on the project due to the vast amount of hours spent on market research, production, design, product testing and marketing. However, we can confirm the value of full range that was donated to the local school who helped us with the design and production process was £12,000.

WHO AND WHAT HAS BENEFITED?

The full Special Needs range is made from 100% HDPE UK post consumer waste, which is more commonly known as milk cartons and the full Special Needs range saved 57,914 two pint milk cartons from going to landfill.

The full range was developed and tested at the end of 2014 and was ready for sale by 2015. Risk assessments were carried out on site and at every stage of the product designs.

Our key outcome and objective was to provide award winning furniture and play equipment that was accessible to all, encouraging inclusive play, working in partnership with SEN Schools and seeking expert advice.

We took into consideration not just fitness and mobility but also behavioural and learning difficulties, as well as emotional health and well being. We worked with the SEN School throughout the design and production process.

The product was tested through our internal rigorous QC process. The design of each product has been to include Special Needs accessibility for all regardless of ability so, where required, ramps have been incorporated to provide access for all and a range of easy accessible seating has been created, thus ensuring the range meets its purpose. The original Special Education Needs range was donated to the local school as a thank you for helping us develop the range. The total value was over £12,000.

Karen Hart, Key Stage 3 Teacher at Villa Real School, said: "Finding outdoor playground equipment for our pupils with learning difficulties or physical disabilities is always frustrating. Discovering Marmax has been a real pleasure as the dedicated staff were so keen to listen to our needs.

"We were consulted regularly on how to produce an optimum design for specialised, accessible equipment. As a result children can play together regardless of their mobility as all models are modified to allow wheelchair users to access them".

Sharon Common, Head Teacher at Villa Real School added: "The kind generosity shown by Marmax Products extended beyond their amazing donation to our school. Tracey and her team took time to work with us to make sure all equipment was modified to our specification. It has been a refreshing change to be involved with a company so dedicated to "getting it right!"

WHAT ARE THE LONG TERM BENEFITS?

The Special Education Needs range encourages active, imaginative play, personal, social and emotional development amongst children of all ages and abilities. The range is practical in design and because it is made from 100% recycled plastic, it is sustainable, robust, durable, and easy to keep clean with no maintenance required. Each piece of play equipment also includes a unique anti-slip surface ensuring the safety of everyone that uses it.

The full SEN Range is maintenance free, which means it does not rot, corrode or splinter and is insect resistant. All products are also resistant to graffiti and more resilient to other types of vandalism. The range is extremely tough and durable with a 10 year construction and UV stability guarantee.

All of Marmax products are manufactured in the UK, comply with BSEN1176 and fall under the rating class 3 in accordance with BS476. They have also been awarded the ISO14001:2004

There were a number of picnic benches in the market place which catered for wheelchair access, but there wasn't any play equipment, made from 100% recycled plastic, which would encourage inclusive play.

The full Special Education Needs range was designed to specifically tackle the lack of accessible play equipment in schools and playgrounds, and provide access for all, no matter what age or ability. The range also met with the government's new initiative for inclusive play.

By installing the Special Education Needs play equipment into the into schools and playgrounds, not only will they be encouraging inclusive play, they are also promoting sustainability by purchasing Marmax products.

WHAT DID YOU LEARN?

We learnt from the project that if one school was struggling to procure outdoor furniture and play equipment for children with a variety of disabilities, there were other schools and industries that would be facing the same challenges. Therefore we developed the complete range and will continue to develop our products to meet market needs.

We will continue to work alongside and listen to requests from Special Education Needs schools to develop new products for our SEN range.

MR C KANNAN

CHENNAI METRO RAIL PROJECT, TAMIL NADU, INDIA

PROJECT AIM

The main aim of the project is during the construction stage is to ensure that the various adverse impacts are mitigated and positives impacts are enhanced. The objectives of the Site Environment Plan during construction stage to be followed by contractors of Chennai metro rail project are as follows:

- To prevent and reduce the negative environmental impacts of the project by implementable mitigation measures, to be carried out by the contractor
- To ensure that the provisions of the Site Environment Plans are strictly followed and implemented by strengthening implementation arrangements
- It is our constant endeavour to take necessary initiatives that our activities are being carried out in such a manner at the impact is not affecting our environment
- Reduce, recycle and reuse of natural and other resources through waste minimisation
- Complying with all applicable Environment legislation and relevant statutory and other requirements
- Periodic Environmental monitoring test will be carried out to ensure all parameters and maintain with in the specifications and necessary and immediate control measures if there are any deviations
- All employees will be given necessary awareness training on environment initiatives and entire site will be motivated to follow the eco-friendly measures and imparted competency enhancement program for environmental team to upgrade with the latest technologies and good environment practices.

PROJECT ACHIEVEMENTS

We at contractors of the Chennai metro rail project are proud to have won safety awards from the National Safety Council of India and the British Safety Council among others or adopting bench marking environment best practices in the construction Industry.

Based on our experience in construction industry, a sound EHS management system is established. This is proactive in nature. It has a broad outlook for EHS management and can handle health and safety problems at its incipient stage.

The EHS department of the company is certified for Integrated Management System (OHSAS 18001:2007, ISO 14001:2004 & ISO 9001:2008). This proves that EHS management system of the contractors of the Chennai metro rail project is as per the international recognition and is compatible with world-class standards and acceptable to global EHS fraternity.

We strongly believe in achieving excellence in safety is a continuous journey and there is always scope for continuous improvement. There is no destination for excellence. Good is not good enough when better is possible and the best is achievable. We always strive to take our safety standards to world class, Global Standards and to international benchmarking level.

WHAT DID THE PROJECT INVOLVE DOING?

The project involves the execution of work activities in such a manner so as to minimise or reduce the environmental degradation and pollution within the standards. A continuous monitoring of environmental parameters like ambient air quality, ground water quality and noise levels is done periodically.

Waste management is given priority importance in the project. Based on the site specific waste disposal plan, we are disposing of different types of waste including hazardous and biomedical waste to authorised agencies. Besides waste disposal, we also promote 3R (reduce, reuse and recycle) practices in our work activities.

WHY DID YOU DO IT?

Development and growth without environmental protection and conservation can never be sustainable and hence it is everyone's duty to perform his/her action in such a manner so as to cause negligible or minimal harm to the environment. As an employee of this organisation I believe that I owe some duties and responsibilities towards the environment. Environmental equity can be achieved only when we all not only understand but also insures the implementation of our duties towards environment.

WHAT DID IT COST AND WHERE DID THE MONEY COME FROM?

Chennai Metro Rail Limited (CMRL) has received an ODA Loan from Japan International Cooperation Agency (JICA) in the amount of JPY 21,751,000,000 on November 21, 2008 toward the part cost of Chennai Metro Rail Project, Phase 1, in Chennai.

This loan arrangement is the first tranche of aJICA ODA loan, which is about 60% of the total estimated cost of JPY 220 billion, in financing Chennai Metro Rail Project Phase I.

CMRL intends to apply a portion of the proceeds of the total loan to payments under the contract for which this Notice of Invitation to Tender is issued.

WHAT DID YOU ACHIEVE IN TERMS OF SUSTAINABLE DEVELOPMENT, ECONOMY, ENVIRONMENT AND/OR EQUITY?

Sustainability is the process suggested to improve the quality of human life within the limitations of global environment. It involves solutions for improving human welfare that does not result in degrading the environment.

The project aims in the achievement of sustainable development with regards to the environmental equity. Through environmental equity we refer development, implementation, and enforcement of environmental policies and laws to ensure that no group or community is made to bear a disproportionate share of the harmful effects of pollution or environmental hazard because of our activities.

Although there is no general agreement about the precise meaning of sustainability, there seems to be a general consensus that three basic concepts are involved in sustainable measures: living within certain limits of the earth's capacity to maintain life, understanding the interconnections among economy, society, and environment, and maintaining a fair distribution of resources and opportunity for this generation and the next.

WHO AND WHAT HAS BENEFITED?

Chennai Metropolis has been growing rapidly and the traffic volumes on the roads have also been increasing enormously. Hence the need for a new rail based rapid transport system and so the Government of Tamil Nadu decided to implement the Chennai Metro Rail Project.

This project aims at providing the people of Chennai with a fast, reliable, convenient, efficient, modern and economical mode of public transport, which is properly integrated with other forms of public and private transport including buses, sub-urban trains and MRTS.

Environmental awareness and its positive culture doesn't benefit any specific section of society or any group. Rather it benefits the whole society because environment is not having local boundaries within states or countries. A large number of trees has been planted in between the course of this project which will provide fresh air, shelter during summers, a good landscape and will help in curbing out pollution. These practices also generate employment to the local people which in turn can improve their economic condition.

WHAT ARE THE LONGER TERM BENEFITS?

The build-up of a positive environment among the employees of this project will be a long term benefit. The promotion of employees' commitment towards environmental issues and needs will make a sustainable environment.

We are conducting a lot of training and awareness to every of our employee and worker as we believe for a positive environmental attitude, knowledge of environment and its component is very necessary. A pro-environmental behavior can be developed by an organisation only when its employees have a positive environmental attitude.

During the execution phase of these contractors of the Chennai metro rail project, we have planted approximately more than 68,000 saplings with a survival rate of 77.17%.

In the coming years these saplings will grow into mature trees and will benefit the society.

WAS THERE ANYTHING INNOVATIVE ABOUT THE PROJECT?

A positive attitude held by the employee towards the organisation and its values. An engaged employee is aware of the environment awareness on management and works with colleagues to improve performance within the job for the benefit of the organisation in relation to the environment.

Awareness programs related to the environment management, protection of natural resources, plantation of trees, global warming are imparted to staff and workmen and the workmen with positive attitude in regards to the environment are selected as torch bearers who will lead the campaign and take it forward with expert reviews from external experts who will review from time to time.

The management also sets its objectives and targets and management programs are devised and taken up on environmental aspects and good practices related to the environment. These include:

- Usage of eco-friendly G3 polymer instead of Bentonite powder for soil stability during piling works
- Usage of fly ash along with cement in construction activities, and usage of fly ash bricks in temporary works
- Usage of excavated soil for back filling and usage of water for sprinkling derived from dewatering wells.
- DG, Crane (Manitowoc) & Compressors sets provided with acoustic enclosures
- Battery operated loco in place of diesel loco
- MoU with approved agencies for disposal of municipal solid waste, bio medical waste, paper waste, plastic Waste, e-waste & hazardous waste
- PUC for vehicles every six months
- Green belt development activities beyond contractual requirements and transplantation of trees and monitoring survival rate.

WHAT ARE THE LONGER TERM BENEFITS?

The contractors of the Chennai metro rail project have signed a Memorandum of Understanding with ITC Limited for the disposal of waste paper, which is a unique national recycling initiative introduced by ITC limited called the WOW (Wealth out of Waste) movement.

Contractors of the Chennai metro rail project joined in this movement and actively participated and contributed to the recycling of paper and saving trees. The movement encourages elimination of waste paper generation at source. Contractors of the Chennai metro rail project provided jumbo bags at all site offices, Stores and adjacent to the Xerox machine and printers. ITC limited appreciated our active involvement and participation and awarded certificate of appreciation for partnering with them in this movement and for continued support in saving trees and making India clean & green.

CAN OTHER ORGANISATIONS BENEFIT FROM IMPLEMENTING YOUR METHODS?

Yes. Other organisations and communities can benefit from implementing our methods as a large number of construction industries are doing the same work and if the techniques and innovations which we are implementing are also used by others, then it will result in the better environmental conditions.

WHAT DID YOU LEARN AND ARE YOU PLANNING ANY FURTHER DEVELOPMENT?

Being a part of this organisation during this project, I have learnt a lot of values which helps in building a positive attitude for environmental protection and conservation. We are providing basic environmental knowledge through various type of training programs to our workers and employees, which makes them to think globally about the environment and gives support on their behalf for eradicating those issues.

We are also promoting environmental friendly practices in our project such as the e banning of plastics on work premises and prohibiting smoking that causes air pollution and bad health effects.

In future development we are looking to increase the scope of this training and environment friendly practices by joining local people also which resides nearby our project area. We can make arrangement for formal education and trainings to general people in parks during the morning hours to understand the importance of a good natural environment around us.

NOMENCA

BARROW WASTEWATER TREATMENT WORKS

PROJECT AIM

Barrow WwTW (Waste Water Treatment Works) in Cumbria serves a population equivalent of 71,772 people. The WwTW was originally commissioned in 1996. The project involved installing a permanent automated dewatering sludge system to replace the existing temporary centrifuge on site.

A centrifuge is a piece of equipment that puts an object in rotation around a fixed axis (spins it in a circle). The dewatering process is the removal of water from solid material or soil by wet classification, centrifugation, filtration, or similar solid-liquid separation processes.

Sludge is the solid matter flushed into the sewage system from the local community's households and businesses. Nomenca delivered the project using SAFE (Skilled Assembly in a Factory Environment) off-site construction.

This innovative approach to construction means that products are produced off site in a safe environment resulting in a reduced programme of works, improved health and safety, guaranteed quality and reduction in wastage of raw materials.

This approach meant that Barrow WwTW Nomenca Limited were able to make significant savings to the project, with the delivery time being reduced by four weeks, a reduction of risks in health and safety (removal of up to 10,000 onsite man hours/annum) and considerable reduction in emissions and environmental impacts.

PROJECT ACHIEVEMENTS

The previous dewatering system at Barrow WwTW comprised a Filter Press Plate but due to performance, environment and health and safety risks, the plant was decommissioned.

Installing the permanent automated dewatering system created positive impacts for health and safety, surrounding wildlife and local community. Nomenca's innovative approach to SAFE off-site construction significant savings across the project were made, including financial, environmental and health and safety. These included:

- Four weeks early completion resulting in a 50% reduction in the time on site
- Reduced health and safety risks on site (removal of >10,000 on site man hours/annum).
- Reduced environmental impacts (1,720kg CO_2 embedded carbon reduction).
- Reduction of emissions and odours which can cause disruption to the local community, may lead to severe health issues and adverse ecological impacts on aquatic and nonaquatic plants and animals
- Capital Expenditure (CAPEX) reduction (14% as a result of offsite construction and assembly, reduced time
- Carbon Footprint – less waste and improved efficiency.
- Equipment tested in factory environment installation on site leading to Right First Time (zero defects).
- Reduced impact on customers by fewer vehicles, less disruption and less time on site.
- Improved reliability and quality.

WHAT DID THE PROJECT INVOLVE?

Nomenca Limited installed a permanent automated dewatering sludge system, replacing the existing temporary centrifuge on site with a turnaround time of 16 weeks and thereby ensuring minimum disruption to the local community.

Nomenca's design team identified a number of cost, timesaving and sustainable innovations that could be provided via 3D Modular Construction, SAFE (Skilled Assembly in a Factory Environment) off-site construction. Products are produced off site in a safe environment resulting in a reduced programme of works, improved health and safety, guaranteed quality and reduction in wastage.

In the pipeline: a new centrifuge system was fitted

The existing dewatering system and filter press control panel were removed and replaced with new controls for the new automated centrifuges and associated equipment. Created via 3D modelling and constructed at Nomenca's Chemical Dosing factory, this initiative resulted in a product with more flexibility and adaptability. The project was completed four weeks earlier with a reduction of more than 30 heavy vehicle movements, thereby minimising disruption to the community.

WHY DID YOU DO IT?

The client required a new automated centrifuge system to efficiently remove sludge from the treatment plant. The water outflow did not comply with the regulatory requirements and so could contaminate the environment. There was also an issue with the associated odour to the surrounding community.

WHAT DID IT COST?

The final cost of the overall scheme was £1.4m and was financed by the client, United Utilities.

WHO AND WHAT HAS BENEFITED?

Sustainability

The previous dewatering system at Barrow WwTW comprised a Filter Press Plate as a fixed volume filter. It is built to hold a specific quantity of solids. Now the system has been updated, it is more sustainable as it is capable of servicing the growing local community.

Economy

Under traditional construction methods, a team would have had to build a platform to house the pumps on site, before a mechanical team assembles the pumps, an electrical team wire them up and a commissioning team tested them. Nomenca did all of this and signed off on quality before transporting the equipment to site, using SAFE off-site construction.

Environmental

With the SAFE process Nomenca reduced potential health and safety risks on site (removal of >10,000 on site man hours/annum). The use of offsite build and assembly techniques, reduced heavy vehicle movements (>30), reduced time on site and reduced carnage needs resulted in a 1,720kg CO_2 embedded carbon reduction.

Carbon footprint reduction

Less waste as once the sludge has been dewatered, you are then left with a cake-like substance which is then given to local farmers for them to utilise on their crops resulting in zero waste. In turn, this meant we reduced the amount of unwanted nitrates and pesticides contaminating the water tables. Local wildlife in the vicinity of the sites were given great consideration and measures were taken to ensure that the project did not interfere with the natural surroundings.

Equity

Nomenca identified a number of cost, timesaving and sustainable innovations that could be provided through 3D

Modelling and our SAFE process. The programme was completed four weeks earlier than planned.

Benefits

The innovation of SAFE off-site construction not only benefited the client financially, once in place the new equipment meant less operator time and maintenance, but also Barrow WwTW and its local community and wildlife by reducing the environmental and health and safety negative impacts.

The Barrow WwTW employees are much safer with minimal handling and operator time on the automated dewatering sludge system rather than the Filter Press Plate.

Carbon footprint reduction due to off-site construction and significantly reduced vehicle movements. The new automated system is more economical for the growing population of the local community.

The local farmers are able to use the recycled sludge cake to fertilise their crops which benefits the local produce, community and reduces contamination of the local water tables.

WHAT WAS INNOVATIVE ABOUT THE PROJECT?

At outline design stage, Nomenca's team identified a number of cost and time saving innovations that could be provided through the use of modular construction called SAFE. The construction intensity was reduced dramatically as so much was completed offsite.

The SAFE approach is currently changing the industry, with major company's realising the efficiencies and savings it can offer.

The SAFE approach provides an opportunity to construct a product with more flexibility and adaptability that supports lower CAPEX (Capital Expenditure) cost. Off-Site modularisation is performed within a workshop environment, offering a greater degree of quality control and assurance, as well as reducing potential health and safety risks associated with site works.

We are very proud of the progress we have made in developing a culture where off site build has become part of projects we undertake. We work proactively with our clients to continue to develop this approach to be safer, more efficient and reduce waste.

Our carbon footprint has been significantly reduced and we have been able to build structures in a safer environment whilst making sure everything is tried, tested and audited before it leaves the factory floor. We believe that other companies and communities would benefit from our methods.

WHAT HAVE YOU LEARNT?

Through this project we have now won additional offsite build projects:
Simester STW Ferric Sulphate Package Plant Scottish Water – £358,800
Llandinam WTW UV Package Plant Severn Trent Water – £272,000
Rodmore WTW UV Package Severn Trent Water – £64,000.

NOMENCA
GREEN APPLE AWARD WINNERS 2015
GREEN APPLE AWARD WINNER 2015
GREEN WORLD AMBASSADOR 2016
HELPING OTHERS TO HELP THE ENVIRONMENT
PEOPLE INSPIRE EXCELLENCE
NORTHMID.CO.UK | PR@NMGROUP.UK.COM

NORTHERN IRELAND WATER

STONEYFORD INTEGRATED CONSTRUCTED WETLAND

PROJECT AIM

In April 2014, NI Water (NIW) broke ground on an innovative new wastewater treatment process – the establishment of its first Integrated Constructed Wetland (ICW). Built for the village of Stoneyford, outside Lisburn, the seven-hectare site is the first municipal wastewater treatment facility of its kind to be created in Northern Ireland.

The objective of the scheme was to provide improved waste water treatment for the village, while creating an aesthetically-pleasing area, rich in biodiversity with potential recreational capacity. The new Stoneyford ICW is a flagship project for NI Water and an industry leading example of how waste water treatment can be integrated into and complement a local indigenous ecosystem.

The £1.3m project included a new pumping station with screening, new pumping main and construction of the new ICW site, which accounted for £800,000 of the total cost. The ICW provides a fully sustainable, low cost, low energy and low maintenance alternative to the modern, high technology-based wastewater treatment solutions normally adopted for a population of almost one thousand people.

PROJECT ACHIEVEMENTS

- Use of local materials and trades with minimal 'external costs'
- Lower capital cost to construct than a conventional aeration process
- The straightforward design minimises construction and maintenance skills required.
- Only technology is monitoring instruments, none of which are critical to the treatment process
- Low maintenance requirements and exceptionally low energy demands
- Ease of commissioning/decommissioning
- Ability to tolerate variable flows and influent quality

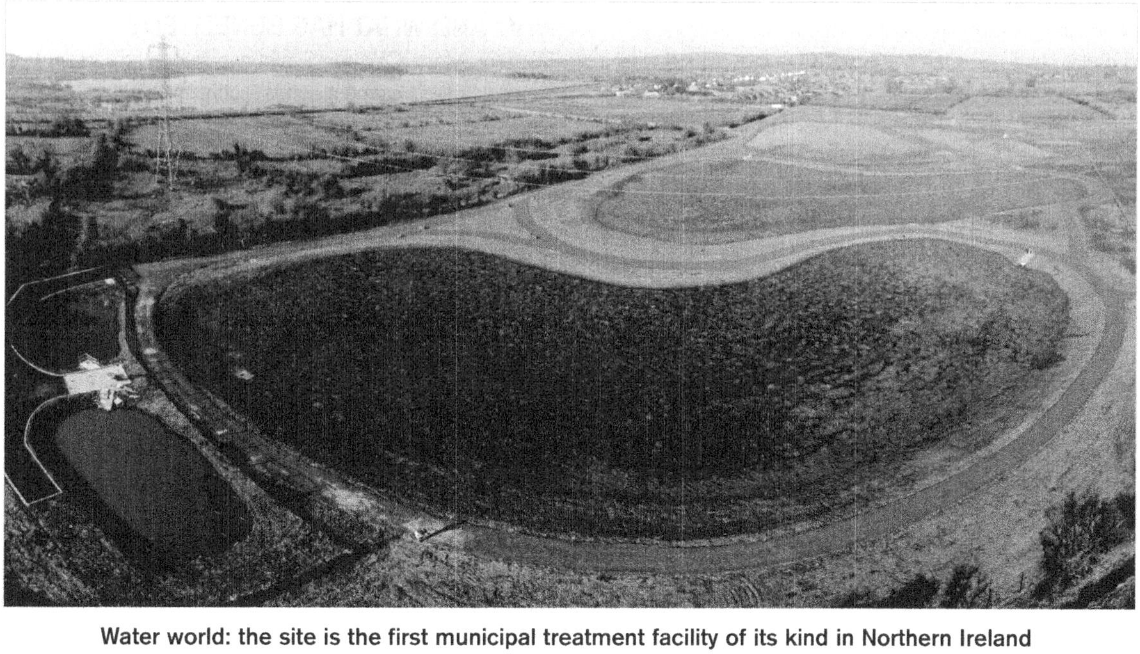

Water world: the site is the first municipal treatment facility of its kind in Northern Ireland

- Exceptionally low sludge production and thereby no sludge tankering
- Sustainable over 50-plus years, a true long-term solution)
- Increased biodiversity and secure large scale wildlife habitats
- Landscape fit and enhanced scenery
- Creation of an aesthetically-enhanced area with amenity potential.

WHAT DID THE PROJECT INVOLVE DOING?

In 2011 NI Water (NIW) sought to upgrade Stoneyford Wastewater Treatment Works (WwTW) which served a population equivalent of 750.

Built in 1983, the existing works – an activated sludge plant process – was overloaded and in the years prior to 2011 it had failed on a number of occasions to meet its Water Order Consent standard for treated water being 20:30:5 (BOD:SS:N). As a result of the poor performance, Planning Service (NI) enforced a development lock until something was done to upgrade the treatment capacity.

The traditional solution was an activated sludge and tertiary plant within the existing site. At this time NIW was searching suitable sites to trial alternative innovative, greener solutions. One of the solutions researched was the use of constructed wetlands, an idea which had been successfully adopted by private businesses as well as a number of councils in the Republic of Ireland.

Often the biggest obstacle facing the construction of a constructed wetland is the amount of land needed – typically in this design at 35m2 per person – but land footprint was less of an issue at Stoneyford due to NIW's ownership of a vast catchment for the water reservoir. NIW engaged VESI Environmental, a company well linked to the ICW process in Ireland. NIW worked with them to assess a number of working sites and found several examples similar to the proposal for Stoneyford village.

WHY DID YOU DO IT?

NIW and VESI derived three key objectives:

- The ICW should preferably be unlined using only materials within the existing site
- The design should integrate the wetland into the landscape and enhance its biological biodiversity through the containment and treatment of influents within emergent vegetated areas
- The aesthetic placement of the constructed wetland into the local landscape should enhance the site's ancillary value and nature management.

Having thoroughly assessed the appropriate location within the village, NIW set about designing their first integrated constructed wetland. A communications campaign was initiated to gain public confidence and importantly to get buy in from the local community on this 'new' wastewater treatment concept.

The proactive two-way communications strategy included a series of public meetings, information events, councillor briefings, leaflet drops and stakeholder visits to other ICWs which provided Stoneyford residents the opportunity to witness first-hand the sensory and social elements associated with a typical ICW site.

Planning permission was granted and the contract for the construction of the project was subsequently awarded with earthworks commencing in May 2014.

WHO AND WHAT HAS BENEFITED?

The Stoneyford ICW consists of two sludge settlement ponds and five treatment ponds – containing over 60,000 wetland plants – each pond having a typical water depth of up to 200mm. The bottom and sides of the ponds have been constructed utilising 30,000 tonnes of suitable on-site soils to provide containment and limit any infiltration to or from ground water

The site is gently sloping from inlet end to outlet, providing the ability to gravitate the flows entirely within the site without any need for inter-stage pumping. All pond embankments are low gradient to the water's edge as a health and safety precaution, as well as providing structural stability and biodiversity enhancement.

The design population equivalent was 952 and the total area of the site is approximately seven hectares, of which 3.7 hectares consist of treatment ponds – the equivalent size of five international football pitches. The ponds are arranged in sequence to maximise the retention time of the influent (wastewater) - design retention time being 90 days - and provide the natural opportunity for effluent breakdown with a resultant high quality discharge.

The ICW has been developed through human technology but is based on natural treatment processes, as it is the plants themselves that support the water treatment process. It is a complex integrated system in which the water, plants, micro-organisms, sun, substrate and air interact to improve water quality.

The Stoneyford ICW design included the reuse of an existing NIW building to house the flow monitoring equipment. The building also facilitates meetings and presentations. The site is already very popular within the industry with group visits from CIEEM, CIWEM, Irish Water, the Institute of Water and the Institution of Civil Engineers.

The site has been landscaped around the ponds to promote its links with nature and incorporates the Stoneyford River and a tributary along the lower boundaries. The proximity of the ICW to the village and adjacent reservoir provides future possibilities for public access for recreation and environmental education. Roadways, access and viewing points have been provided to allow safe and easy access for pedestrians.

Initial flows to the ICW commenced late November 2014. NIW is monitoring the performance at each pond and to date is comfortably meeting the Northern Ireland Environment Agency's (NIEA) discharge licence limits of bio-chemical oxygen demand (BOD) 15mg/l, suspended solids (SS) 25mg/l and ammonia-N 3mg/l.

Other more mature ICW systems of similar size are achieving nutrient removal rates of up to 99% for parameters such as BOD, suspended solids and phosphorus, as well as pathogen removal (faecal and total coliform) to almost non-detectable limits.

CAN OTHER ORGANISATIONS BENEFIT?

NIEA has been very supportive of the Stoneyford ICW project and like NIW believes that this natural environmentally-friendly solution is particularly well suited to rural areas. NIW aims to introduce more of these sustainable solutions in the future with planning applications already submitted for ICWs at Castle Archdale and Ballykelly.

"We had gone to Glaslough ICW expecting to see marshy bogs, swarming with flies and wreaking of a foul smell. What we actually experienced was completely different. There was no smell at all – even at the inlet ponds – and hardly a fly in sight. Groups of pony trekkers were enjoying the lush green amenity space and we even caught a glimpse of a Moorhen darting around the ponds. What surprised us the most however, was how clear the water leaving the final ICW pond was – it looked clean enough to drink!"
Stoneyford resident on an NIW organised informational site visit to a similar ICW constructed in the Republic of Ireland.

NOTVITCO

THE ORDINARY SKINCARE COMPANY

PROJECT AIM

We set out to build a business to help those who suffer from dry and sensitive skin, employing these key elements:

- We formulate our products using sustainably sourced high quality materials
- We will not use wild farmed ingredients
- We will not use any raw materials that have statistically insignificant evidence supporting their efficacy
- We only use recyclable packaging
- Our packaging must enable 99%-plus of packed product to be dispensed, avoiding product waste inherent in some systems
- We outsource our manufacturing to a very high quality contract manufacturer compliant to ISO 22716.

PROJECT ACHIEVEMENTS

- In the first year of trading we were a finalist in Marie Claire Prix de Excellence de la Beaute Awards 2015, so we were placed in the top three from 135 entrants
- We have launched six retail products compliant with our over-arching objectives, and three more for the professional market
- We have grown a network of stockists in South Africa, and launched on-line businesses there and in the UK.

WHAT DID THE PROJECT INVOLVE?

The Ordinary Skincare Co project started with identifying a need for effective healthy skin moisturisers. There are a plethora of products available but most contain additives that are meant to enhance the skin but frequently trigger reactions or worse, have long term deleterious effects.

We developed moisturisers with a bare minimum of ingredients. Extensive market research and knowledge of the anatomy, physiology and science behind skin and skincare was of critical importance.

We decided to manufacture in Cape Town. South Africa is a multi-ethnic society with some of the harshest climates for skin. Unemployment is high and small business plays a crucial social and economic role in the developing economy. All manufacturing and packaging was conceived to be safe for humans and the environment.

WHY DID YOU DO IT?

The idea started with observations Claire made during 25 years in medical practice. She saw many women who developed skin problems. She thought they were related to ageing and hormonal changes, but realised the common denominator was the use of anti-ageing skin products.

She researched the effects of "anti-ageing" additives while conducting a study for a transdermal hormone product. Some skin cream ingredients can lead to thinning and long term degradation. These ingredients were often described as natural. She realised how many different substances we were exposing our skin to routinely, some of which are "natural botanicals" like tobacco!

Research showed about 50% of people report having dry and sensitive skin or eczema. Few products alleviate dry itchy skin without triggering sensitivity. Sometimes a product would be identified that didn't trigger responses but it didn't moisturise well either.

One of Claire's children became seriously ill, and her skin suffered terribly as a consequence. One day she asked: "Can't you just find an plain ordinary skin cream for me""? We knew we couldn't find one having searched all our lives but maybe we could make one? We knew we had the name now – The Ordinary Skincare Co.

The key factors in designing our formulae were:

- Effective moisturising
- Minimum ingredient inventory
- Tried and tested ingredients with a successful heritage
- Modern ingredients of proven efficacy
- Minimal sensitisation.

Years of formulating and testing followed. Using cosmetic safety database resources we identified the most effective ingredients, with environmental safety credentials. They had to be sustainably farmed and essential for efficacy. Even when considering fragrance, we screened them for the mildest natural ingredients.

Making products to the highest standards is vital. Robert and his vast experience in the chemical industry was invaluable. Our products are:

- Made to ISO 24716
- Approved for sale in the EU, the best international standards for cosmetic manufacture.

We saw other companies proliferating their ranges pointlessly. We knew a focussed simply effective range was what was required and also we needed:

- Eye-catching packaging
- Great dispensing systems that stored product safely, dispensed everything, and worked well everywhere
- Dispensing systems and packaging that were fully recyclable.

As part of our business model, we knew we didn't want a big company with a workforce. We planned to outsource everything possible and operate using modern systems to even obviate the need for fixed offices, including:

- Websites were created to handle on line sales linked to fulfilment services to handle pick-pack-deliver in the UK (for the world) and South Africa. They were built to the latest standards by a handpicked UK firm
- Back-office processes linked into the websites
- Sales via Amazon in the UK to extend reach.

All of the investment required has been from our own resources. We used our retirement savings to fund the business. We have been frugal but effective.

After the first year of operation we achieved the following:

- Growing brand recognition in all markets served
- Media recognition including radio interviews and feature articles in national magazines
- Third finalist in the SA Marie Claire 2015 Prix 'Excellance de Beaute Awards
- SA Sunday Times Lifestyle Top Three South African beauty products
- Developing network of loyal stockists in South Africa and Increasing number of repeat buyers
- Positive feedback from clients everywhere
- Clear proof those with problem skin now have a reliable solution
- Proof that our ethical and straightforward "bricks free" business model can be successful.

WHO AND WHAT HAS BENEFITED?

Our main beneficiaries have been our customers, many of whom use Ordinary Skincare as part of their skin care regime. Their feedback shows we have really "got it right".

Suppliers benefit because we spread our business over a number of small efficient local enterprises both in South Africa and the UK.

WHAT ARE THE LONGER TERM BENEFITS?

We see the business becoming a permanent part of the industry. The signs are all encouraging. Three charities are supported via our website. On check-out, customers vote for their choice and we make a donation in proportion to their votes.

Our strong points are:

- We are leading a small but steady revolution in attitudes to skin care. We believe in offering products that simply work, without hyperbole, and provide comfort to those who suffer troublesome skin conditions
- Our innovation has been to understand the medicine and science behind sufferers' problems – and provide effective solutions for all types of skin
- Our products have been designed to avoid anything unnecessary. There are no inventively described or exotic ingredients of dubious benefit
- We apply credible scientific method to the business and ensure everything is thought through carefully, with the aim being to ensure all aspects of the products and service is meticulous. All without compromising on the glamour of a beauty product
- We have applied our strong environmental awareness and "eco ethics" to every aspect of the business
- We have a simple approach to all aspects of the business. We have developed each element of the business carefully, and tirelessly worked to achieve a high standard.

O'BRIEN CONTRACTORS LTD

ENVIRONMENTAL IMPROVEMENT

PROJECT AIM

At £200m the Nestlé group's project at Tutbury is their largest construction project worldwide, and will consolidate all their UK coffee production across the UK. Sisk were appointed as main contractor in 2012 and O'Brien Contractors won a £12m groundworks and civil engineering package to prepare the site for the major construction project.

Our specific challenge was that the wettest year on record (2012) had eroded the suitability of the Nestlé original engineer's design solution to groundworks to facilitate the preparatory work on site – the site is situated in the middle of the flood plain of the River Dove and Mill Fleam. The effect of this change was circa an additional £250,000 and three weeks on site, pushing up price and programme before we had even started the job.

The design required O'Brien's vehicles to pass through Nestlé's existing factory complex on their roads, through their own traffic and employee's pedestrian routes. Specified works alone required at least 9,000 20-ton lorry movements. Over 40,500m3 of spoil were to be taken off site and transported to landfill as it was not suitable for re-use, with an estimated 270,000 miles of haulage (16.5T CO_2).

PROJECT ACHIEVEMENTS

The results of our intervention with an innovative alternative solution were:

- 7,000 fewer lorry journeys resulting in the saving of 15 tonnes of CO_2
- Saving fuel for 210,000 miles of 20 tonne lorry haulage
- voiding over 40,000m3 of spoil going to landfill
- Reduction of the imported stone by 70%
- Avoiding disturbing birds feeding on the wetlands
- Avoiding a three week minimum programme extension
- Avoiding 9000 lorry movements across an operational factory site and goods haulage roads/yards
- Saving £800,000
- Avoiding inevitable damage to existing site roads
- Creating a stable base for building 23,000m2 of roads and service yards, plus 20,000m2 of building footprints on a development site of 124,000m2

O'Brien Contractors measure customer feedback on their work, during and after the project's completion. These are used to create KPIs and allow individual projects to be benchmarked against each other to find our best practice.

Sisk also scored us six monthly on industry KPIs against 1,864 other subcontractors in their supply chain. The system scores 1 to 5 (Integer only) for 4 topics and our scores were:

- Health & Safety – 4
- Quality – 4
- Commercial – 4
- Programme & Performance – 5.

With an average KPI of 91% we were placed 12th out of 1,865 subcontractors.

WHAT DID THE PROJECT INVOLVE?

Our directors held a brain-storming workshop to consider possible solutions and one emerged that was considered very suitable, uncommon and most likely would not have occurred to the designers when they planned the works much earlier in 2012, but the changing water conditions now made it an ideal option.

O'Brien Contractors put forward the option of using lime stabilisation to avoid taking the spoil off site, create stable conditions for alternative site road access and the working hard standings for the construction works and stable conditions for buildings, roads and yards required in the final development while reducing risks, costs and programme.

After our internal brain-storming workshop we spoke to specialist subcontractor TR Stabilisation for their collaborative early input and worked up the feasibility study to present to Sisk.

This was then presented to Nestlé's team leader, Faithful+Gould and presented both the problem and a solution, with indicative costs and programme implications. Having demonstrated to Faithful+Gould the implications and benefits of our innovative proposal, we were then asked to meet with Nestlé and explain matters to them, which included the reductions we were making in potential safety issues.

Key aspects of the O'Brien innovation were:

- Significantly reduced construction transport across Nestlé operational areas
- Reduced lorry movements by circa 7,000
- No spoil to landfill
- Created safer stable site for ongoing construction work such as craneage
- Savings of £800,000
- No programme delay instead of three weeks-plus minimum with original scheme
- Significant increased risk accrued to O'Brien, requiring doubling of their PI insurance.

Our risk assessments demonstrated our innovative proposed lime stabilisation solution had no more significant risks than the original scheme at site level and significantly reduced many other risks.

WHO AND WHAT HAS BENEFITED?

Social

The biggest social impact our innovative solution delivered was preventing 7,000 20-ton lorry journeys through the centre of Tutbury and the surrounding residential areas. Preventing this number of vehicle movements across Nestlé's existing site significantly reduced the probability of an injury incident occurring to the joint work forces.

Environmental

Our lime stabilisation solution delivered two significant environmental benefits. Not taking 40,500+m3 of spoil off site and avoiding it going to landfill avoided significant environmental impact. The change to on site stabilisation meant a net saving of 7000 20-ton lorry trips, which equated to around 15 tons of CO_2 saved by preventing 210,000 haulage miles. These changes meant that herons and swans feeding on the wetlands were not disturbed.

Economic

Our solution reduced costs and time. The value of this element of the works as tendered was £1.75m and additional works required would have raised this to £2m. Our innovative solution came in at £1.2m, so we saved £0.8m.

In time terms there were at least another three weeks required for the original specified works plus effects of deteriorated ground, which would have had knock-on economic effect on following works packages as well. The snow and thaw in January would have made matters worse, so we prevented at least three weeks delay. Our solution will have longer term economic benefit for Nestlé too as there was no damage to their existing site roads and the new site will last longer with lower future maintenance.

WHAT WAS INNOVATIVE ABOUT THE PROJECT?

We believe that our best practice approach to finding an innovative and safer solution on this project has been the central enabler to the new relationship we have established both with Sisk and Nestlé. We are currently working on two further schemes with Sisk – a £2.2m contract at Number One Eastside Locks, Birmingham and a £1.6m contract at Lyons Park, Coventry.

Our positive and beneficial safety impact on the wider community is unquestionable. Our supporting information shows that the limited width A511 road, which accesses the site through Tutbury, and the unmanned railway level crossing adjacent to the site entrance meant that more than 7,000 lorry loads would have had to cross to access the A38 if we had not changed the scheme.

The resulting traffic congestion and wear and tear of the roads and level crossing would have increased the potential for accidents. The busy pedestrian crossing of the Nestlé bell mouth road entrance on route to the train station would also have been at possible risk.

For our own team the way we approached this project has been an exemplary lesson in the innovation and value we are able to bring to a project, as well as the reductions in environmental and safety impacts that our value engineering can deliver as a by-product of saving time and money. It has made us more confident in our own abilities.

CAN OTHER ORGANISATIONS BENEFIT?

The solution we used on this project is repeatable elsewhere, where ground conditions make its use suitable. It is particularly suitable for developments such as Nestlé's which lays central to a major flood plain between rivers.

With global warming effects and the increase of flooding in the UK, we believe it will become a more valuable option for clients and designers to add to their site feasibility studies.

We plan to provide a case study to organisations like the CIOB and Midlands Best Practice Clubs to offer them case study presentation events in conjunction with TR Stabilisation to make this solution more widely known to organisations who are not civil engineers.

Compared with corresponding 2012 Constructing Excellence KPIs our performance is:

- Mean AIR – Industry = 536
- % All Contractors/Subcontractors achieving 0 AIR = 65%
- O'Brien Contractors' Project Dove AIR = 0

We are in the top quartile on performance.

@ONE ALLIANCE

CAMBRIDGE GROWTH SCHEME, WATER RECYCLING

PROJECT AIM

The project was to construct a new section of a water treatment works in Cambridge to allow for future growth of the city. The aim was to complete the project within budget whilst ensuring minimum disturbance to the local community and having as little impact on the environment as possible.

PROJECT ACHIEVEMENTS

The site team was able to complete works on site within 10 months. We exceeded our waste diverted from landfill, embedded carbon and potable water usage targets, and all excavated material was kept on site for reuse and a number of innovations were utilised during the project which reduced time on site and our overall carbon footprint for the job. For example:

- Roadcem (binds with material on site to make a structurally sound base for a road)
- Weholite (100% recycled plastic pipework and tanks)
- Above ground plastic pipework (rather than below ground concrete pipes)

WHAT DID THE PROJECT INVOLVE DOING?

In 2014 the Cambridge growth scheme project started. This £20.9m project involved constructing a new extension to the existing water treatment facility. The new works were required in order to handle the forecasted population growth in the region and to replace ageing infrastructure on site.

This included the construction of an activated sludge plant which was four lanes wide (in total 100m by 36m); an activated sludge plant pumping station; four final settlement tanks; a sludge thickening area; and 1,400m of above ground pipework.

The client was Anglian Water and they set a number of environmental targets including;

- 80% waste diverted from landfill
- 50% reduction in embodied carbon
- % potable water usage.

Careful planning of the works meant that all of these targets were exceeded. In addition, a number of best practice methods were implemented during construction, and are highlighted below.

WHY DID YOU DO IT?

As a responsible company, Anglian Water has environmental targets that are set annually and our alliance is responsible for helping to deliver these. In response to this, the alliance has worked with the client to develop a sustainability roadmap which looks to achieve these goals.

Our main aim was to ensure industrialised construction, which involves off-site prefabricated and pre-engineering technologies, was taking place. Speed was one of the biggest advantages of using this construction method, however it did also provide a more sustainable project with the added benefit of being a financially smart solution.

WAS THERE ANYTHING INNOVATIVE ABOUT THE PROJECT?

Road stabilisation

Rather than conventional construction methods for the haul roads that involved importing thousands of tonnes of stone to the site, the construction team decided to opt for a more sustainable alternative called Roadcem.

This is a fine powdered blend that enabled the binding of the existing ground to form a structurally sound road surface. By using this product, we avoided importing 11,500 tonnes of virgin stone to site, we avoided any haulage off site, and the road hardening could be left in place to create a final access road for the client.

Pipes and tanks

The project had approximately 1,400m of cross site pipework to install, linking one area of the treatment works to another. The project team engineered a no dig solution for the installation of the pipework, meaning that the vast majority of infrastructure was installed above ground, eliminating approximately 1,300 tons of excavated waste.

Traditional construction methods would have used either concrete or steel pipes, however, we used a product called Weholite, which is made from 100% recycled plastic. This material was also used for the construction of a large modular tank which was manufactured off-site.

By doing it this way, it improved the operational efficiency on site, had a better design life and was cheaper to manufacture than concrete pipes.

Material

In the initial survey for the site, it was identified that the surrounding area to the development required the levels raising for access, and site drainage. In order to reuse the 36,000 tons of subsoil and topsoil we had created, we applied for a materials management plan to allow this to happen.

By retaining the vast amount of material on site, we ensured that we would achieve our waste diverted from landfill target and reduced vehicle movements off-site by 1,500 wagons. This in turn meant less fuel consumption and less traffic nuisance through the city centre.

Pre-cast concrete

For the construction of the activated sludge plant and final settlement tanks, we used prefabricated and pre-cast concrete sections that were manufactured off site. To join each section, a small channel of concrete was poured to secure them in place.

By taking this approach, we had a 2% concrete wastage figure versus an industry standard 6%, because of the off site pre fabrication. Overall, this reduced the construction time by approximately 40%.

Potable water

Pressure testing is required of all new installations of pipes and tanks, this process usually requires large amounts of potable water. As an alternative to using potable water, the construction team utilised the final effluent from the existing wastewater process. Over the project lifetime, this equated to 11,870m3 of saved potable water.

Financial and other benefits

The innovations used on site and mentioned above had financial benefits as well. These included:

- A saving of £93,600 by reusing our 36,000 tons of material on site.

Ins and outs of the project: internal and external view of re-cast concrete ASP lanes

- A 65% reduction in the embedded carbon footprint. This reduction came from the materials used and construction processes on site
- A saving of 11,870m3 in potable water equating to approximately £720 in water from the client
- £3,380 saving of waste removal costs due to a no dig solution for installing pipes.

WHO AND WHAT BENEFITED?

- We spent less time on site which reduced the health and safety risk to employees, there was less disturbance to the local community meaning happier customers, and financial savings for the client and the alliance
- The surrounding communities now have a newly-refurbished water treatment works up and running, and it's one that can deal with the forecasted population growth over the next ten years.
- Through efficient project planning, we managed to reduce fuel consumption, embodied and operational carbon, water usage and waste off site, meaning overall less traffic movement and disturbance through heavily urban parts of Cambridge.

WHAT DID YOU LEARN FROM THE PROJECT AND ARE YOU PLANNING ANY FURTHER DEVELOPMENT?

From start to finish, this £20.9m project took 10 months to complete. This was an enormous accomplishment for the site team.

We also exceeded the target of diverting 80% of construction waste from landfill (99%) and are working towards achieving our first 'zero waste' project.

Upfront planning is absolutely key to succeed in the project delivery. Early engagement with all involved parties ensures that the outcome is a sustainable one.

OSTLERS CIDER MILL

CIDER VINEGAR

PROJECT AIM

To be recognised for my green efforts and to be certified organic (currently under conversion).

Ostlers Cider Mill is one of a few traditional family-run cider orchards in Devon that grows and hand picks its own apples to make cider and vinegar.

Ostlers' Cloudy Apple Cider Vinegar with Mother is unpasteurised and has a growing UK wide reputation both to use in cooking as well as for its health benefits.

PROJECT ACHIEVEMENTS

- Gold awards at Taste of The West and Great Taste
- Product used on finals week of BBC2's Great British Menus
- Our product has been written about in The Ethicurean Cookbook and talked about on LBC radio

WHO AND WHAT HAS BENEFITED?

Environmental

The management of existing four acres of orchard, four acres dedicated to wild life and the planting of a further three areas of new orchard

Innovation

Managing my orchards with a mixture of horses and sheep which will organically fertilise the land and this way we can keep all of the wild flora and fauna without turning it to turf.

Social

By creating a unique product 'vinegar with the mother' which is unavailable in supermarkets and can only be bought in local independent shops.

Economic

By managing the land and business with our own natural resources, animals for the land and solar and wood for the house and office.

By installing a solar system for water and electricity and by using animals instead of sprays and fertilizers and wood burner for heating and water.

ARE YOU PLANNING ANY FURTHER DEVELOPMENTS?

- To have a certified organic product and to replace some of the machinery used in the process of apple picking by hose and cart.
- Commitment to achieve by senior personnel involved
- Stopping the use of sprays and fertilisers, registering the land as organic, installing solar. We now use our apprentices fathers' business to recycle any waste we cannot reuse from packaging.

CAN OTHER COMPANIES BENEFIT?

We believe our style of land management could be used by others. The bonus is that we are very self-sufficient.

OXFORD BROOKES UNIVERSITY

LOW CARBON OXFORD

PROJECT AIM

Contributing to creating a low carbon Oxford is an ongoing challenge. With 2,800 staff and 18,500 students, Brookes needs cleaner, greener and sustainable means of getting people between its eleven Oxford sites.

Cleaner and greener: the BROOKESbus is embracing new technology

Since 2004, the BROOKESbus has provided the solution. The fleet has moved almost 12 million passengers, since 2009, saving an estimated 1.4 million car journeys each year. The BROOKESbus is essential to continue to reduce the university's transport carbon emissions, which account for approximately two-thirds of the total University emissions.

This project aims to further reduce carbon emissions on existing BROOKESbus trips whilst encouraging more staff and students to swap car journeys for bus ones.

PROJECT ACHIEVEMENTS

In 2014 the 20 buses of the Brookes fleet become the first commercial buses in the country to be installed with the Gyro-drive System.

The Kinetic Energy Recovery System, developed in Oxfordshire by the Williams Formula One team with the support of an Oxford Brookes University alumnus, "harvests" energy into a flywheel during breaking. When the bus accelerates, energy is fed into an electric traction motor.

The BROOKESBus network travels over 520,000 miles per year and the upgrade to Euro 6 vehicles and fitment of the Gyro-drive System has allowed the network to reduce its CO_2 emissions by 17% against the previous contract, saving an estimated 68,500 litres of fuel and 876 tonnes of CO_2 per annum.

NOx emissions have also been drastically reduced by the Euro 6 technology, with reductions of as much as 95% against a Euro 5 vehicle. This has helped to improve air quality in Oxford city centre, which saw 58 breaches of the hourly mean objective for Nitrogen Dioxide in 2012 and 12 breaches in 2013. Following the introduction of the Oxford Low Emission Zone (LEZ) on January 1 2014, no breaches occurred during the course of 2014.

WHAT DID THE PROJECT INVOLVE DOING?

Oxford Brookes' Travel Plan strategy has been in place since 1999 and aims to foster a culture of sustainable travel and reduce transport related carbon emissions and dependence on the private car.

The BROOKESbus service, first launched in 2004, is critical to achieving these aims and connecting the university campuses, student halls and the city.

WHY DID YOU DO IT?

The BROOKESbus is a five-year project between Oxford Brookes University and the Oxford Bus Company with a £2.2 million yearly investment made by the university. Profits are split between Oxford Brookes University and Oxford Bus Company.

Since 2009 the service has been run in partnership with the Oxford Bus Company (The Go-Ahead Group) and moved nearly 12 million passengers. This service saves an estimated 1.4 million car journeys a year. As part of our commitment to sustainable travel, we sought to introduce a system that would reduce the carbon emissions of this service.

In 2014 all 20 buses of the Brookes fleet were installed with the Gyro-drive System, resulting in up to 17.9% fuel saving and a 17% reduction in CO_2 emissions. The technology is the first in the country to be installed on a commercial bus services and has been developed by our alumni working at Williams Formula One. Previously, it has been used in racing cars including vehicles competing in the Le Mans 24 hours endurance race.

These were the first Euro 6 rated vehicles to go into service in Oxford city, and featured superfast 4G Wi-Fi and e-leather seating. This project combines a number of firsts:

- The first commercial deployment of flywheel technology to service buses in the UK
- The first examples of the ADL Enviro 400 MMC brought into commercial service
- The first Euro 6 vehicles brought into service in Oxford.

There is no legal requirement to fit the Gyro-drive System but the university specified this in order to further reduce its environmental impact. As a result, we were able to create a project that successful combined these various innovations in a high-quality, appealing package that delivers for the customer day in, day out. It is not only our staff and students who benefit from this service. The BROOKESbus network also provides essential transport routes to the community, including key routes for schools neighbouring our sites.

WHO AND WHAT HAS BENEFITED?

The BROOKESbus project successfully embeds economic, social and environmental benefits. By increasing the number of students using public transport, the scheme significantly reduced parking and traffic problems in and around the university campuses, whilst providing transport routes to the community and improving sustainability.

Positive impacts of the scheme include the following:

- The 50% increase in students using the BROOKESbus as their main commuting option and the 40% decrease in students using cars between 2009 and 2014.
- The services saves an estimated 1.4 million car journeys a year
- Enhancing student experience with free travel and improved connectivity between the campuses, student halls, city centre and the train station.

ARE YOU PLANNING ANY FURTHER DEVELOPMENTS?

We are eager to build on the success of this scheme and sought to do so by reducing the emissions of the BROOKESbus fleet. The introduction of the new Gyro-drive System is estimated to be saving 68,500 litres of fuel and 876 tonnes of CO_2 per annum across the 20 buses in the fleet.

Compared with traditional battery based hybrid systems, this mechanism performs at lower costs and weight and moreover, it requires minimal re-engineering.

The new Euro 6 buses save up to 95% of the harmful NOx emission against a Euro 5 vehicle, helping to improve air quality in Oxford and reducing the possibility of breaches of the hourly mean objective for Nitrogen Dioxide. We are working closely with Oxford Bus Company to monitor these figures and ensure that we are making the most reductions possible.

So far, the service has surpassed the model shift targets set against the service which was 30% of usage and reached 50% in 2014. The project also resulted in excellent positive PR for Oxford Bus Company, Brookes and their partners, with the story being covered in BBC South Today and the local press, as well as featuring widely across the trade press.

45 PALL MALL VENTURES

45 PALL MALL

PROJECT AIM

45 Pall Mall is a recent instruction (Q4 2014) for JLL. When JLL took over the management we identified that there was no joined up approach to the waste regime at 45 Pall Mall. This was apparent at our first occupiers meeting (Q4 2014). After listening to our occupiers we have instigated on-site training for the on-site cleaning staff and the on-site security teams on how to monitor the waste and to give advice to the occupiers and their own cleaning staff.

The aim of our project was to review and begin establishing a market leading waste and recycling process at 45 Pall Mall. This would take place in a number of ways, but primarily via refresher training, auditing and collaboration with key stakeholders such as occupiers and relevant service partners. Our end goal is to achieve 70% recycled by the end of 2015, with new streams being trialled within the building, such as food recycling, clothing and batteries.

We also wanted to remove the use of black bags from within the waste streams and to establish cost savings on the overall waste management budget via improved recycling figures, a reduction in waste coming out of the building and finally, via rebates for materials such as cardboard.

PROJECT ACHIEVEMENTS

When JLL took over the management we instigated training for the on-site cleaning staff and the on-site security teams on how to monitor the waste and to give advice to the tenants' own cleaning staff on how to dispose of in the correct waste.

The building manager underwent a training course and was successful in gaining the NEBOSH award in Environmental Awareness at Work.

The occupiers now receive colour coded sell-through bags as part of their service charge. Office waste is now separated into recyclable and non-recyclable streams. As a result, the waste collection company now only has to collect twice daily.

This has improved the waste recyclable figures from 45% to 50% (average for 2014) to between 65% and 75% and the resultant 50% decrease in the emissions from the waste trucks.

There are no loose bags being left on the pavement and has resulted in an improved look outside 45 Pall Mall and the collection trucks now use service roads to the rear of the building.

WHAT DID THE PROJECT INVOLVE DOING?

45 Pall Mall is located on the northern side of Pall Mall, one of St James's most prestigious addresses. The building is on the doorstep of St James's Park and Green Park and benefits from excellent local amenities and transport connections.

Owned by M1 Real Estate is a contemporary office building in an enviable location in one of London's most refined quarters. The building provides around 60,000 square feet of contemporary accommodation with excellent natural light and benefits from an expansive remodelled stylish reception with retained finishes.

WHY DID YOU DO IT?

Before the current Facilities Management team began on site in late 2014, waste management had not been high on the agenda, but the incoming team saw an opportunity for an improved scheme which would benefit the companies in the building in terms of not only cost savings, but also the desire to show corporate social responsibility.

The team at 45 Pall Mall, their service partners and the occupiers are now committed to providing, as far as practicable, an economic and sustainable waste management approach resulting in a less costly use of economic and environmental resources.

Recycling of all waste where possible is an essential part of the sustainable waste management system at 45 Pall Mall. The existing system is based on two main streams — mixed recyclables and general waste, which accurately covers the majority of the needs of the building.

However, a recycling average percentage of 50% in the previous Q4 of 2014 highlighted that whilst the streams may be suitable, the implementation was not recording data which we believe promotes a sustainable ethos.

Working with our cleaning and waste service partners, we commissioned a full building audit to review all parts of the recycling process, from source to collection. This involved collecting waste from the building and taking it to a specialist sorting facility where the waste streams were analysed for their content — as were users who might be recycling better than others.

The findings were in line with our beliefs in that there was plenty of scope to not only recycle more but also to reduce the amounts coming out of the building entirely, a key focus of any waste management strategy. We took three main points from this audit:

- Around 30% of waste currently going into our general waste (non-recyclable) stream can actually be recycled
- The use of black bin bags was noted for general waste correctly, but also used for both recyclables and cardboard streams. The use of black bags automatically marks waste as non-recyclable
- 75% of our recycled stream, is made up of paper and cardboard, with very little contamination, which is excellent.

WHO AND WHAT HAS BENEFITED?

As a result, we now had factual evidence with which to engage our occupiers and our service partners and detailed to them not only the highlights of the findings, but also how 45 Pall Mall facilities management team and the occupiers would look to address these in partnership with the buildings waste company, cleaning and the occupiers own cleaning companies.

This included the following key points:

- Removal of all black bin liners for the building. This will allow all cleaning teams to continue their good work of segregating recyclable and non-recyclable waste bags
- The installing of common area Euro bins (11,00 litres)
- The cutting down of 5x collections a week to 3x collections and no waste to be lefi for collection on the pavement area of Pall Mall
- The renewed signage in bin locations
- Trial of food recycling with a view to further increase recycled amounts coming out of 45 Pall Mall
- Relocation of large bins (Euro 11,00 litres) within loading bay/car park area, ensuring correct bins are used for waste and allow for better management
- Recycling roadshows based in our reception areas. These will engage with your staff to become more aware of waste and recycling and the streams we have in place in the building
- Roll-out of refresher training of the on-site cleaning teams, just to update on the new methods and to ensure they buy into the vision, recognising the key role that they play
- Installation of a bike storage area. During Q1 2015 an under-used area of the lower ground common area was converted into a cycle secure storage area, which allows our occupiers for the first time to enjoy the facility of a secured covered parking area for 18 cycles.

WHAT ARE THE LONG TERM BENEFITS?

Resulting from these recommendations, positive changes have begun occurring in a short time space, with the reduction of black bags being used, the relocating of bins, the re-education of cleaning staff and tenants' cleaning staff within the building, and the forming of a Green Forum within the occupiers.

This has seen our recycling percentage up 10% in the period February to April of 2014, which is a fantastic result.

Many of the measures which have been employed in the building have been delivered at no cost. This has come about by simply by giving the subject the attention it deserves and engaging with our teams and tenants.

Costs will be associated in the roll-out of food recycling, but our aim is that this stream is fed from the general waste stream, and so with collection costs at a similar rate, there should be no major cost increase.

PETERBOROUGH HIGHWAYS (SKANSKA)

ONE PLANET LIVING

PROJECT AIM

The aim of the project was to deliver a highways maintenance contract in a sustainable way. Typically the industry is inherently not sustainable or environmentally friendly. The Peterborough Highways contract wanted to show that environmental improvements could be made in the industry and that there was a great deal of room for innovation.

The client (Peterborough City Council (PCC)), also has the ambition of becoming the UK Environment Capital and delivering the contract in this way supports PCC in this endeavour, thereby reducing the impact of essential road maintenance.

PROJECT ACHIEVEMENTS

The project has achieved a number of objectives across the One Planet Living theme, including:

- 97% of waste was diverted from landfill in the first year of the contract
- Installation of solar PV panels generating on average 40% of the energy for the building per month
- Use of less CO_2 intensive working methods such as the "Dragon Patcher" and micro-asphalt heaters
- Fuel savings through fleet innovation and choosing an electric car over a conventional fuel vehicle
- Increased local procurement across the contract.

Green wheel: electric car made fuel savings

SKANSKA

WHAT DID THE PROJECT INVOLVE DOING?

In 2013 Skanska signed a ten-year contract with Peterborough City Council for the provision of highway maintenance services. As one of the industry's leading green contractors, Skanska is using its Peterborough Highway Services partnership with the council to model its use of green technologies and sustainable techniques.

This chimes with the council's own commitment to become the UK's Environment Capital – leading in all areas of the environment, from green spaces to sustainable development.

Determined to make Peterborough Highway Services truly sustainable from the outset, we developed our contract implementation plan around the sustainable One Planet Living framework, which is a global initiative developed by the World Wildlife Fund for Nature (WWF) and Bio-regional.

We created a suite of added value KPIs, which measures performance against the One Planet Living themes. This is used to monitor and manage our performance and communicate performance to elected members, as well as operational teams.

Our approach has also received the Overall Contribution to the Environment Capital award at the annual Peterborough Green Awards. Judges noted our achievements in integrating the principles of the One Planet Living strategy into the implementation of Peterborough Highway Services, diverting waste from landfill and community engagement projects.

WHY DID YOU DO IT?

By its nature, construction has the potential to create negative impacts on the natural environment when not

Cost cutting: the Dragon Patcher (left) and a micro-asphalt heater helped to eliminate waste

actively managed. Skanska sees a big opportunity in proactive environmental management and indeed, it is Skanska's ambition to become the global leading green contractor.

Peterborough City council are committed to becoming the UK Environment Capital and therefore provided client support for Skanska's 'green' ambitions with both parties benefiting from the environmental improvements made.

Financial

Our initiatives have generated savings such as reducing the amount of waste to landfill (saving transportation and landfill tax costs) and reducing fuel consumption

- £7,000 pa saving on electricity (from photovoltaics)
- £3,600 pa saving on fuel costs through using GPS system
- £650 in fuel costs pa through using the electric vehicle
- 150m3 pa saving/£219 pa in water through rainwater harvesting/grey water.

Savings are being returned to the council through an efficiency mechanism in the contract, which ensures the client retains best value. We are also adding value by working with other organisations:

- Sustrans – the sustainable transport charity gave a cycle tour of Peterborough to help us better understand the importance of good design in developing cycleways. This will benefit our client and the Peterborough community as a whole
- PECT – we have sponsored the Eco Schools Awards, which recognises schools for their environmental initiatives
- Peterborough DNA - we are working with this innovation program, part-funding an environmental graduate placement on Peterborough Highway Services.

WHO AND WHAT HAS BENEFITED?

In the first year, we exceeded the target of diverting 95% of waste from landfill (97%), and are looking to increase this. Of the materials procured for Peterborough Highway Services, 55% have come from the LEP, which helps to minimise travel and therefore carbon emissions. We are working with suppliers to increase this to 80%.

We are making sure that food procured for our office canteen is produced locally and sustainably where possible, including having Fair Trade accreditation. Currently 60% of non-materials suppliers (labour, consultancy, etc) are from the local area, outdoing the original 50% target.

In addition, more than half of our Peterborough Highway Services staff are helping to develop community skills, through supporting the activities of the local Skills Service, a job and training brokerage funded by the Greater Cambridge Greater Peterborough Enterprise Partnership (LEP). We have also recruited apprentices and are working with Through The Gate, a scheme to reintroduce ex-offenders into the workplace.

We are working towards achieving a zero carbon economy with an annual reduction target of 5% of direct emissions. Having collected a data set over the first 12 months of Peterborough Highway Services, it was apparent that 55% of our direct carbon emissions came from the diesel we used for fleet vehicles and business journeys. We are now using a GPS system in all of our fleet to encourage efficient fuel consumption and better driving techniques.

Elsewhere, we are reducing the carbon emissions from the energy use at our contract office by fitting solar photovoltaic panels. This has replaced around 40% of the building's total energy consumption. Baseline studies highlighted that half of the potable water we used related to vehicle wash-down.

We now use waste wash-down water from underground storage tanks that pre-wet the road salt as part of our gritting activity. Also, rainwater harvesting will be installed for use within the vehicle wash-down process and gully cleaning.

WHAT WAS INNOVATIVE ABOUT THE PROJECT?

One innovation has been our use of the TomTom GPS system in our Euro 6 fleet and company cars. This is helping our people to drive in a more environmentally-responsible way and helps us identify the greenest drivers.

To help us achieve our target of zero waste, we are implementing alternative technologies which minimise waste in the first place. These include the Dragon Patcher, a viable alternative to more wasteful and costly cut and fill repairs, and a micro-asphalt heater which also provides a rapid solution to potholes.

To minimise carbon emissions, we have allocated a fully electric vehicle (Nissan Leaf) and an electric bicycle to undertake the short trips taken on the network. The car is covering 7,700 miles per year (5,130 over eight months so far), saving around 950kg CO_2 emissions.

WHAT HAVE YOU LEARNT?

The benefits of the Peterborough Highways contract will be felt across both the organisation and by the wider Peterborough community.

Directly

- Short/long term cost savings to the client and best practice that can be rolled out across the wider Skanska business, generating further savings
- Increased health and wellbeing for employees
- Helping deliver Peterborough's goal of becoming UK Environment Capital.

Indirectly

- A gradual greening of the Peterborough area
- As part of the Smart Fengate working group, we are able to help small local businesses prosper through our local procurement strategy, by providing access to internal resources for HSQE advice/support, tendering help for small companies and the use of meeting room facilities.

The key lesson learnt from this project was the need for early engagement with the client, in this case during the bid process.

By ensuring the understanding of the client's aspirations it is possible to develop a contract that can mutually deliver both parties requirements in an appropriate fashion.

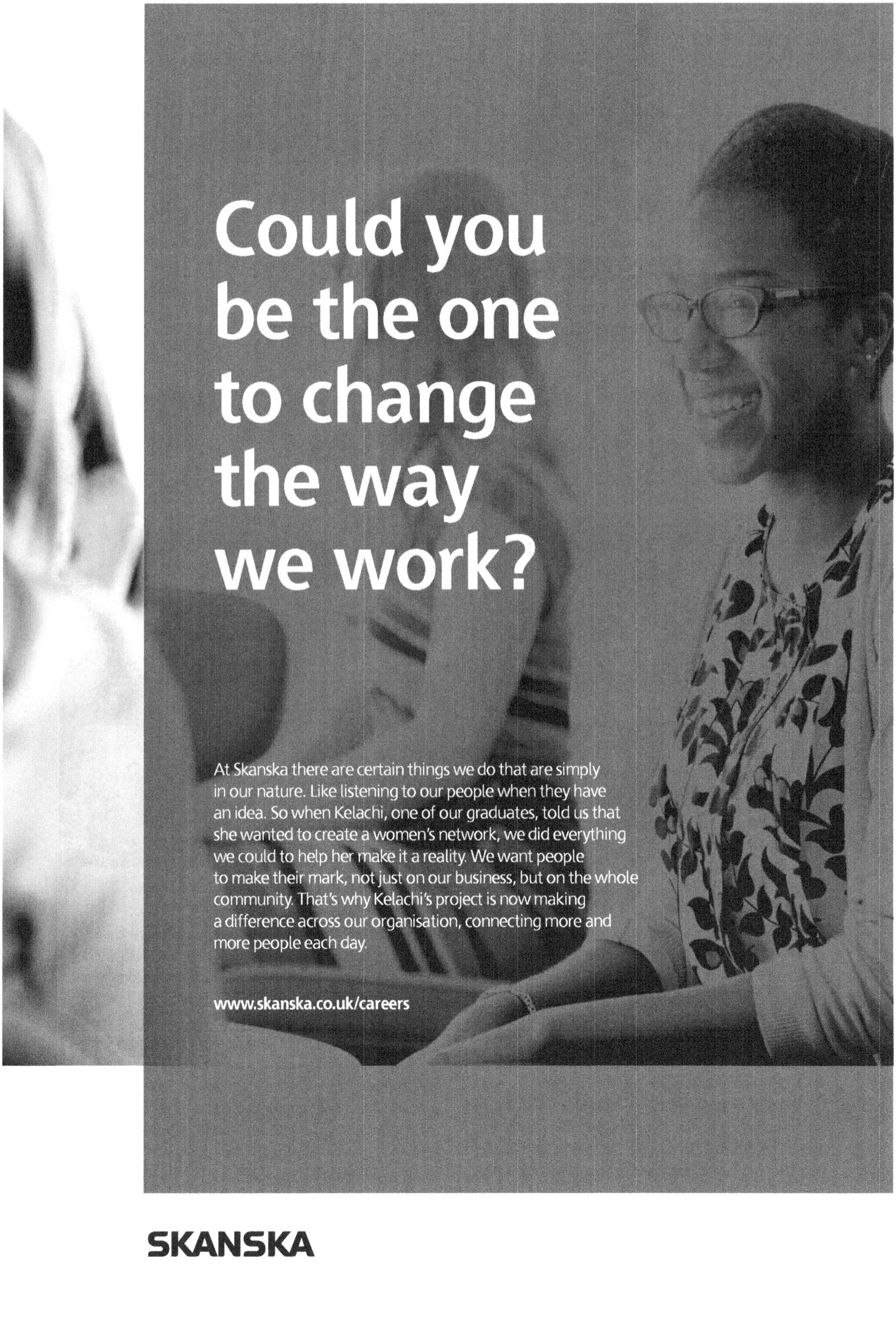
Could you be the one to change the way we work?
At Skanska there are certain things we do that are simply in our nature. Like listening to our people when they have an idea. So when Kelachi, one of our graduates, told us that she wanted to create a women's network, we did everything we could to help her make it a reality. We want people to make their mark, not just on our business, but on the whole community. That's why Kelachi's project is now making a difference across our organisation, connecting more and more people each day.
www.skanska.co.uk/careers
SKANSKA

PROJECT JANSZOON, NEW ZEALAND

TRANSFORMING THE ABEL TASMAN NATIONAL PARK

PROJECT AIM

Project Janszoon is a privately funded environmental trust undertaking the ecological transformation of New Zealand's 22,000 hectare Abel Tasman National Park, in partnership with the Department of Conservation (DOC) and the community.

Cutting through: the clean-up has got rid of predators such as rats

The Abel Tasman is one of the country's most popular national parks with 230,000 visitors annually.
Biodiversity values across the area were compromised prior to its formation as a national Park in 1942 due to farming and logging operations, and subsequently by the invasion of weed species and introduced predators including rats, possums, stoats and wasps.

In 2011 philanthropists Neil and Annette Plowman agreed to support Project Janszoon – named after Abel Janszoon Tasman, the first European to discover New Zealand in 1642. The project is budgeting to invest $25m over the 30 year period to 2042, the 100th anniversary of the park.

Project Janszoon was established to implement a new concept of collaboration with DOC, whereby philanthropists commit to fund an ecological transformation of a part of the conservation estate against agreed and defined indicators, and the government commits to secure those ecological gains, once achieved, for the long term.

PROJECT ACHIEVEMENTS

Since its launch in 2011 Janszoon has made significant headway with extensive predator control, replanting projects and weed control programmes under way. Bird reintroduction has commenced and an education programme, smart phone app and Wi Fi network has been launched to engage the community and visitors in the restoration process.

In 2015 Project Janszoon was honoured to receive both the Philanthropy and Partnership, and Supreme Award at New Zealand's Green Ribbon Awards. Less than four years into this unique partnership the building blocks for significant environmental change are in place, its work is being recognised at the highest levels, and the project is becoming a blueprint for further philanthropic partnerships.

Winging it: the clean-up has seen the return of many species of bird

WHAT DID THE PROJECT INVOLVE DOING?

The Abel Tasman is New Zealand's smallest national park, at 22,530 hectares, but it is also one of the most popular with 228,000 visitors annually. Visitors are attracted to the beautiful coastline and the four-day Abel Tasman Coast Track walk.

There are a wide variety of habitats in the park, from sub-alpine tops to sandspits and islands, and more than 70 bird species have been recorded. However many of those birds are now only found in small numbers in the upper reaches of the park, or have gone from the area entirely.

Project Janszoon Director Devon McLean had spent many decades visiting the park and watched as birds like robin and kakariki, and trees like flowering rata and mistletoe, that were once commonly seen along the coastline, disappeared.

A history of fires and land clearance by both Maoris and Europeans, and the introduction of predators like rats, possums, stoats and wasps had significantly diminished the Park's biodiversity values.

With limited funding, the Department of Conservation (DOC) was forced to prioritise its spending and had made the decision to primarily fund visitor assets in the park.

An opportunity to reverse the ecological decline was identified and Project Janszoon was formed in 2011 with a plan to invest $25 million over 30 years. The project aims to provide a safe environment for the return and rejuvenation of flora and fauna by greatly reducing predator and weed numbers by the 100th anniversary of the national park in 2042.

The vision is to also nurture the passion of the community, school children, visitors and to ensure the gains made are preserved by future generations. There was no blue-print on how to undertake an ecological restoration in a national park and a lot of work has gone into building relationships between Project Janszoon, DOC, the Abel Tasman Birdsong Trust which was already working in the area, adjacent land owners, iwi and concessionaires who make their living in the park.

WHY DID YOU DO IT?

Initially a strategic plan was commissioned which became a framework and reference document for the project. DOC established a technical advisory group and Project Janszoon assembled its own expert team and also established an external Advisory Board to challenge the thinking and guide initial stages.

WHO AND WHAT HAS BENEFITED?

The approach to ecological transformation involved three overlapping phases. In the first three years much of the focus was on the SECURE phase to remove negative influences.

Predator control required significant investment in infrastructure and 15,000ha (70%) of the park is now tracked and equipped with stoat traps. In August 2014 an aerial control operation for rats was completed in response to heavy seeding from the beech forests in the park leading to an explosion of rat numbers.

Control work has also progressed against goats, deer and feral cats in areas where restoration plantings are occurring. Monitoring shows these methods have already made the park a much more secure habitat for birds, animals and plants.

A weed survey found 113 different ecologically damaging weed species. The project has supported the Abel Tasman Birdsong Trust's wilding conifer and weed removal projects and committed funding to ongoing conifer control. Wasp eradication trials have confirmed the effectiveness of this work in reducing the competition between these insects and the nectar feeding birds.

With extensive predator control in place the second phase of work to RESTORE lost biodiversity and revitalise ecosystems began in 2014.

An aviary was constructed in the upper reaches of the Park to permit soft release of yellow crowned kakariki which commenced in 2014. These birds are the progeny of wild captured birds taken from nearby predator free islands and raised at four specialist aviaries in the upper South Island.

The project's ornithologist led the work for the Birdsong Trust's release of saddleback on Adele Island in the Park in 2014. We are attempting to attract seabirds, including shearwater and gannets, back to the coastline using sound systems and decoys with unconfirmed reports of gannets landing in one breeding area. Planning is presently under way to return South Island robin, kaka, tuatara, brown teal, blue duck, and kiwi to the park.

A planting project has begun to restore one of the district's last remnants of lowland kahikatea forest at Hadfield Clearing that has seen 16,000 plants established so far. Seed collection and propagation is proceeding for 125,000 plants to be established over a ten-year period.

Dune restoration at Anchorage is underway as is a programme to "fire smart" much of the foreshore by planting less flammable species to minimise the risk of accidental fire arising from the many park visitors. Replanting of flowering rata along the coast has also begun.

Monitoring of pests, weeds, bird abundance and the condition of indigenous biodiversity as a way to plan control operations and provide performance measures is undertaken regularly.

The third key phase of the programme is to FUTURE PROOF the gains by developing strong community, iwi, and business involvement.

In 2014 the Adopt a Section programme launched. Three pilot schools committed to building a five-year relationship with "their piece" of the park. Students, supported by their teachers and experts from Janszoon and DOC, are learning about their location, studying key aspects of the ecosystem, and starting to take action to aid and monitor the process of restoration.

The education programme is a new way of getting young people to engage with the park and become the environmentalists of the future. Research had found that most schools in the region did not have an ongoing relationship with the park, other than school camp, which emphasised recreation rather than environmental themes.

Students engage in hands on learning about ecology and conservation in relevant and engaging contexts. Some biology, geography and horticulture classes are engaging in research that earns credits for New Zealand's National Certificate in Education Achievement (NCEA). It is planned to introduce more schools to the programme over time.

A Student Advisory Board was introduced with representatives from each partner school to nurture and develop youth leadership.

The goal is to nurture stewardship and the concept of kaitiakitanga (guardianship and conservation) in our young people, so they will protect the park in the future.

A private, solar powered, WIFI network has been built in the park and a free smart phone app that acts as a "virtual visitor centre" developed. The app is another way to educate, raise awareness and interact with people about the park using technology, especially 18 to 35 year olds who make up 70% of park visitors. This unique partnership with DOC is working well with Janszoon funding DOC staff to carry out work like maintaining stoat traps, managing contractors and working with students. Janszoon experts also work alongside DOC staff to achieve the shared restoration vision. In Janszoon's 2014 annual report DOC's Martin Rodd outlined why he believes the partnership is working.

"When you join DOC you want to make a difference and it feels like we are really part of something significant," he said. "We are able to work on projects and priorities together. Project Janszoon brings expertise to the table that coupled with our own experience and skills means we are making ecological gains that we could never have achieved on our own."

As the project has a 30-year time frame Project Janszoon wanted to ensure the government committed to securing the ecological gains it makes past its lifespan. In 2014 Project Janszoon became the first, and so far only qualifying project, to sign the Tomorrow Accord with the government. This pioneering initiative secures the conservation achievements of Project Janszoon, and ultimately the work of other like-minded philanthropists, for the long-term benefit of New Zealand.

WHAT ARE THE LONGER TERM BENEFITS?

The project is also committed to funding research and development into innovative forms of predator control and restoration. In 2012 the first permanent trial of 200 Goodnature A24 self resetting rat traps began across 100 ha at Pitt Head. These traps can reset 24 times reducing the labour required to monitor the network.

With the support of the Birdsong Trust a lot of work has been done to finesse the traps and the latest results are encouraging and may allow the return of the south island robin to the mainland. Trials in conjunction with Zero Invasive Predators (ZIP) Ltd of new stoat lures are also showing promising results. This research will allow new methods of predator control to be used throughout NZ.

There is national interest from DOC in how the WIFI network could help with real time bio-diversity monitoring and the smart phone app's ability to tell stories of our national parks.

Never before has a partnership between philanthropists and DOC been done on this scale and DOC is looking at Project Janszoon as a trial ground for new ideas about how to collaborate in the conservation estate.

The collaborative model that has resulted is held up by DOC as an example of best practice and department heads have visited with their Australian counterparts to get a better understanding of what makes the project work so well and how it could be replicated.

CAN OTHER ORGANISATIONS BENEFIT?

Project Janszoon is contacted regularly by other philanthropic and environmental groups and have hosted visitors, including those from the United Arab Emirates and Australia. Some of these groups have been contemplating philanthropic collaborations of their own and have been encouraged to see the positive results that are already occurring in this partnership between a private trust and DOC.

The Tomorrow Accord was also designed to be used by other philanthropic groups, so they too can secure ecological gains past a project's time span. It is hoped that the existence of this document will encourage more philanthropists to invest in the conservation estate.

Project Janszoon's work in the Abel Tasman National Park would not be possible without the collaboration with DOC and other stakeholders.

A lot of effort is put into these relationships. Project Janszoon works successfully alongside the Abel Tasman Birdsong Trust, who were already undertaking small-scale ecological work in the park. Janszoon has supported some of their existing weed and wilding pine control work and make experts available to help with projects like re-introducing saddleback to Adele Island and funding trapping trials.

As a new and credible voice in the space Project Janszoon has helped work through some historical difficulties between DOC and neighbouring landowners. Along with DOC we spent considerable time consulting with interested parties regarding the use of aerial 1080 and in December 2013 Tasman District Council granted a 10-year resource consent without public notification.

Concessionaires who run tourism operations in the Park are also consulted on many aspects of our work. This, from Wilsons Abel Tasman CEO Darryl Wilson: "The park operators and visitors are already seeing changes in the park for the better. Project Janszoon is communicating a real sense of hope and determination that things are planned and will happen. We are hearing about the dream and being inspired by it, that is important."

WHAT ARE THE LONG TERM BENEFITS?

Project Janszoon's work will ultimately benefit the ecology of the area, the community, and have an economic benefit for the Nelson/Tasman area.

Visitors to the park are currently attracted by the beautiful beaches and scenery and the visitor experience will ultimately be enhanced by frequent encounters with wildlife, and a thriving park environment. Increased tourist numbers will benefit the region and local businesses. Locals are also likely to increase their involvement with the Park.

Project Janszoon's 30 year project will be completed on the Park's 100-year anniversary and 400-year anniversary of Abel Tasman's "discovery" of New Zealand. By 2042 the vision is that:

- Biodiversity values in the Abel Tasman National Park will no longer be threatened by incursions of invasive weeds and pests
- Populations of key indicator species of birds, animals and plants will show favourable trends on all vital measures
- The restoration will be future proofed with strong community, iwi and business involvement
- The model created by the Project will be replicated on other parts of New Zealand's public conservation land.

Nearly four years in we are well along the path to the 2042 anniversary being a celebration of an outstanding ecological success story.

RAININ INSTRUMENT, USA

TERRARACK

PROJECT AIM

TerraRack was created to provide researchers with pristine quality pipette tips in the most ecological manner.

TerraRack will transform the Life Science industry by greatly decreasing its contribution of plastic waste in landfills and fundamentally keeping ecological practices as a priority.

PROJECT ACHIEVEMENTS

Through state-of-the-art design, Rainin has engineered the first thermo-formed tip rack, substantially minimising the amount of plastic required to produce a pipette tip rack.

TerraRacks are 50 per cent lighter than a conventional tip rack and are highly compressible when empty and completely recyclable.

TerraRack is the only lightweight, stand-alone pipette tip rack that can easily stand up to the rigors and forces of pipetting. In addition to being completely recyclable, TerraRack offers an even greener solution by reducing the amount of plastic entering the waste stream.

TerraRack has fundamentally changed the way future pipette tip racks are designed, keeping environmental concerns top of mind. TerraRack is expected to dramatically reduce the amounts of plastic waste produced by life science research laboratories worldwide.

WHAT DID THE PROJECT INVOLVE DOING?

Rainin's new TerraRack pipette tip line minimises plastic tip rack waste and solves a major problem in labs around the world: what to do with used plastic racks?

Spent conventional racks can consume a large volume of space and represent a significant amount of plastic waste. Where disposal rates are based on weight, they are also a considerable expense.

The TerraRack shell, with its hinged lid and front latch, is made of PETE (polyethylene terephthalate, also called PET in some regions), an exceptionally strong polyester that is easy to recycle and lightweight.

A polypropylene tip deck snaps into the shell, adding structural support with minimal weight. Each single use disposable rack is 50 percent lighter than conventional tip racks, highly compressible when empty and completely recyclable.

WHY DID YOU DO IT?

TerraRack has a broad range of features that are highly relevant to life science researchers who are looking for

TerraRack family: a fundamental change in the way pipette tip racks are designed

quality products, optimal performance and convenience. These include:

- Single-use (non-refillable) and pre-sterilised disposable racks
- Sterile tips available both with and without filters
- Shrink-sealed
- High clarity PETE thermo-form for superior product visibility
- Product-specific information printed onto each rack
- Rainin part number
- Product description
- Lot code.

TerraRack is available in both LTS and universal tips in all popular tip sizes (10 µL, 250/200 µL, 300 µL and 1000/1200 µL), including filter, non-filter and ShaftGard. TerraRack design features include:

- Accommodates both single- and multi-channel pipettes, and partial multi-channel pipette loading (the edge of the rack will not interfere with the pipettes nozzles)
- Lightweight, highly recyclable thermoformed PETE shell, the same material used in beverage bottles
- Roughly half the weight of injection-moulded racks
- Hinged-rack design and locking front latch prevents tips from spilling
- 96 tips per rack – ten or eight racks per box
- Not refillable or autoclavable.

TerraRack accessory

TerraRacks are designed to be a sturdy, stand-alone tip racks that will function perfectly right out of the box. For those who prefer a weightier feel, there's TerraBase, an effective and inexpensive option to gain the practical and environmental benefits of TerraRack, while retaining the weight and feel of a conventional rack.

- Two bases: TerraBase Short for shorter tip racks (10 µL to 300 µL tips) and TerraBase Tall for taller racks (1000/1200 µL tips)
- Made with durable, chemical-resistant, autoclavable polycarbonate
- Four silicone feet for non-slip traction on the bench surface
- Completely optional: adds weight and traction for customers who prefer a conventional rack feel.

How TerraRack operates

Like most pipette tip racks, TerraRacks are simple to use. First, remove the protective shrink wrap covering the rack. Second, place the rack on the lab bench, unlatch and open the hinged lid – begin use.

For those that are using the TerraBase, after removing the shrink wrap covering, TerraRack inserts into one of the two bases. Second, unlatch and open the hinged lid – begin use.

TerraRack vs conventional Pipette Tip Racks

In every tip size TerraRack is approximately half the weight of a comparable conventional injection-moulded tip rack. In addition to the PETE shell, TerraRack employs a uniquely designed polypropylene tip deck that also contributes significantly to the reduced weight. PETE is not a suitable material for the tip decks.

RG GROUP

DORRIDGE DEVELOPMENT

PROJECT AIM

The RG Group were appointed as main contractor on a new retail development at Dorridge in the West Midlands. The overall aim was to help the regeneration of the town centre and we took this project as an opportunity to set the following further aims:-

- Minimise waste – reduce demolition waste created and maintain low carbon footprint in doing so
- Trial hybrid generators
- Utilise rainwater harvesting
- Trial waterless urinals
- Enhance biodiversity in local area
- Deliver carbon neutral project
- Contribute to the local community and leave it enriched
- From the aims develop proposals going forward to roll out on other projects across company and share best practice.

PROJECT ACHIEVEMENTS

At the end of the project substantial community and sustainability achievements were met.

- In total 78 tonnes of waste were prevented by reuse within the local community (saving 4.5% of demolition waste and preventing waste to landfill rising by 335%)
- 99.9% of waste diverted from landfill
- Demo waste 1,718.60, construction waste 313.05, Total waste (c,d & e) 16,558.78
- Landfill 17.89 tonnes.
- Waterless urinals saved 140 litres of water per week
- Rainwater harvesting saved 3,000 litres of water
- Hybrid generators – average of 45% of energy supplied
- CCS Score of 45 – 9 in each section
- Reward scheme – 500 cards issues, 70 completed which equated to £700 spent in local shops
- New biodiversity area created in key urban area
- School tree planting programme – 1,800 trees planted.

We have learnt lessons and initiatives to take onto our upcoming projects. The area of Dorridge has been transformed and new life put back into the heart of the town centre. Hopefully the new trees in Dorridge community park will be enjoyed by many for years to come, and the school tree planting programme will inspire children to consider sustainability issues throughout their lives and help to reduce carbon.

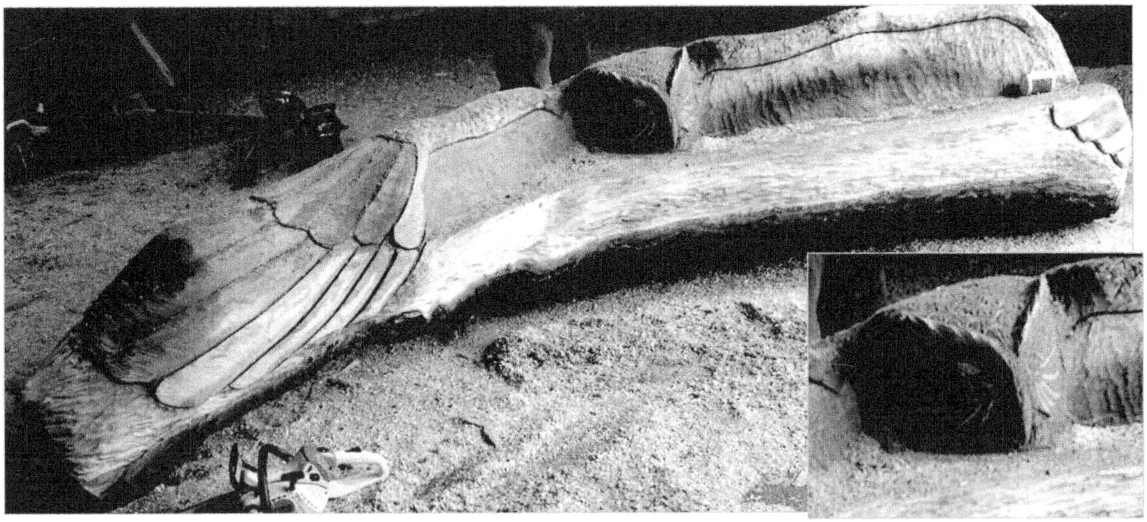

Bird's rest: an oak tree was removed and refashioned into an owl seat

WHAT DID THE PROJECT INVOLVE DOING?

RG were appointed main contractor on a new retail development at Dorridge, West Midlands. We took this project as an opportunity to see how much we could minimise waste being produced, trial sustainable solutions, carry out community and biodiversity initiatives and upon project completion, feel like we had left the community enriched.

The site in the heart of the village centre was previously an old abandoned shopping centre and the new development was a positive key step to the regeneration of Dorridge. As well as a new shopping facility, the development provided for the existing doctors' surgery to be extended and fully refurbished,an independent pharmacy, new parking, a café and business units and associated S278 works.

The projects ran for 42 weeks from January 2014 to November 2014, and had a value of £13.4 million.

We started by carrying out a pre-demolition audit of the site and identified materials which could be salvaged and reused. As we wanted to involve the local community in every step of the project, we then aimed to try to find local homes for the materials.

A press release was put in the local paper and members of the public were invited to see if they could reuse the available materials. At the same time we had an on-site Environmental Champion, who was contacting local community schemes to see how we could help them.

The materials identified for reuse were: slabs, pavers, timber, logs, security spikes, di-bond hoarding, railings, barriers, paving, kerbs, racking, noticeboards, containers, water boiler, cabling, traffic cones, furniture, demolition arisings, muck away, green waste, mulch, bike hoops, and surplus tiles.

All of these items found homes within the local community and went to various sports clubs, local halls, a cricket club, bowling club, village hall, local residents.

WHO AND WHAT HAS BENEFITED?

A main part of the demolition was the existing shopping centre steel structure, which was dismantled and a home

Roots to success: tree-planting at a local school

was found for it at the local Tyseley Locomotive Works as a cover for vintage trains.

The project set up an off-site consolidation centre where reverse logistics were used and the materials available for reuse were stored to prevent damage prior to being utilised. All waste was able to be segregated here as well allowing for a higher recycling rate from any final waste.

All new materials were sent via the consolidation centre which allow for JIT deliveries to site, thereby preventing any damage and waste arisings and resulted in reduced transport into the village and reduced resulting traffic and carbon arisings.

To further reduce waste in the greater sense of the term, several sustainable initiatives were trialled, these being hybrid generators to power the site cabins, waterless urinals and rainwater harvesting. As a result of the project we have now implemented waterless urinals across all our projects, do rainwater harvesting where feasible and are doing a further trial of new solar hybrid cabins.

The Environmental Champion on site was tasked with monitoring and collating all carbon data from the project. In conjunction with Carbon Footprint and Tree Appeal, RG offset the embodied carbon of the project, making it a carbon neutral project. As well as a certified VCS offset in India for a wind power project, we planted native British species trees, all grown in England from collected seed.

Each UK tree accounted for one tonne of carbon offset resulting in 1,888 trees being planted primarily within schools. A tree planting day was held at a local school and also with the council at the Dorridge community park where mature oak trees were planted.

As part of lowering the impact of the project on the community and environment, parking was not permitted within the village centre and off- site parking was arranged. There was a permanent information centre set up in the high street and a resident drop in day/evening and news letters organised.

A workforce recognition card was introduced to encourage the use of local businesses by the site employees, and rewards were given for best practise on site for environment, community and H&S and then the rewards could be spent locally.

A volunteer day at a key local RSPB Reserve – Sandwell Valley – was organised where a team helped to create feeding habitat for wading birds and a runway for nesting birds. A donation was also made towards habitat regeneration at the reserve.

Local schools were engaged with informative visits on careers within the construction industry and, via the British Beekeepers Association, we introduced an Adopt a Beehive scheme at five local primary schools.

As part of the works there was a large oak tree that had to be removed. Not wanting this to go to waste we commissioned an artist to turn the tree into an owl bench, which was sited near to the new surgery for waiting visitors to sit and enjoy.

Throughout the project the site team engaged with local community initiatives such as the refurbishment of Bentley Heath Community Centre, helping in local events and supporting local charities.

Most of the initiatives employed on this project cost very little to implement with the exception of the carbon offsetting and were part of a projects standard budget. Costs were in terms of man hours which mainly were on a voluntary basis. The community will benefit for years to come and we believe the initiatives can be easily adopted by other contractors.

ROCHDALE EXCHANGE SHOPPING CENTRE

CAR PARK LIGHTING EFFICIENCY

PROJECT AIM

We aimed to improve the appearance of the multi-storey car park facilities through the introduction of LED lighting and an ANPR ticketless system, to help make the car park appear more attractive, future-proof and reduce running costs where possible.

PROJECT ACHIEVEMENTS

The newly-installed LED lighting has achieved a reduction in energy consumption, which we have reduced further by closing off the higher parking levels which are used less frequently. We are not lighting areas which are not in use. The multi-storey car park has achieved the Park Mark accreditation and has also achieved the Disabled Parking Award, demonstrating the quality of car parking facilities that we offer.

WHAT DID THE PROJECT INVOLVE DOING?

This project involved the replacement of lighting on a like-for-like basis to put LED fittings in-place of the older and less efficient light fittings. Consultation meetings were held to discuss and demonstrate the ability to increase the lighting levels while still reducing the energy consumption.

The ANPR system was also replaced on a like-for-like basis after considering what equipment we wanted and the locations of the equipment. We held many consultation meetings with the ANPR company to 'build' the specification of what we wanted the system to do, what we expected it to look like and what we could use the system for to help with other site requirements.

We completed this project with a view to reducing all costs where possible to ensure we were maximising revenue for the landlord. At the same time we wanted to increase the appeal to customers by making the car park more attractive and safer to use, and also to reduce our carbon footprint where possible.

WHAT DID IT COST?

The initial outlay for this project was £180,000 which was paid for by the landlord after the manufacturers stated that the replacement lighting had a pay-back period of seven years based on current usage. This is expected to be earlier as we have reduced consumption further through restricting access to the higher levels of the car park, unless current levels are reaching full capacity.

WHO AND WHAT HAS BENEFITED?

Since the introduction of the LED lighting, customers have given us constant positive feedback, stating that they feel the car park is certainly much more appealing and is a safe environment for them to use throughout the year. The landlord has also benefited from a reduction in operational costs.

The ANPR system has helped reduce costs even further as £7,000 a year was being spent on tickets for the previous system which is now being saved. There also is the added benefit that we are not using any paper tickets, underpinning the centre's green culture on site to reduce the impact on the environment in any way possible.

On-site staff are also now able to assist customers much more effectively through use of Smart phone applications.

Due to the nature of the LED light fittings, their life-expectancy is far greater when compared to the previous fluorescent light fittings. The installation was also under-pinned by a fully-comprehensive five-year guarantee, which ensures that the annual maintenance costs are kept to a minimum during the payback period.

The ANPR system is future-proof in that all updates are simply programmed into the system through remote-access, rather than having to make structural/equipment changes.

WHAT WAS INNOVATIVE ABOUT THE PROJECT?

In terms of innovation, there was an extensive use of a combination of motion-detection and light sensors connected with timers throughout the car park to control lighting in areas used less frequently and to ensure the lighting is not consuming energy when the centre is not operational.

With regards to the ANPR car park equipment, a web-based app was created to give on-site staff the ability to assist customers such as enquiries about length of stay and to provide assistance with payment at exit barrier etc. All of this is fully auditable and provides an extra tool for us to use to ensure the customer has the most positive experience possible.

CAN OTHER CENTRES BENEFIT?

We have spoken with a large number of sites regarding the ANPR system as one of the benefits is that it is easily adapted to suit a wide variety of sites, and can be adapted to suit a site's changing requirements even after the initial installation.

We have easily demonstrated to each of the sites enquiring about the ANPR system the massive benefits of using this system compared to the systems that require the use of tickets etc.

WHAT HAVE YOU LEARNT?

The introduction of LED lighting has resulted in significant operational savings and as a consequence, we are now considering extending the technology into the malls within the shopping centre to reduce the service charge.

ROYAL PRIORS SHOPPING CENTRE, LEAMINGTON SPA

WASTE MANAGEMENT INITIATIVE

PROJECT AIM

The Royal Priors Shopping Centre in Royal Leamington Spa has already achieved zero landfill disposal status through contractor waste services utilising waste to energy plant. In order to minimise the effect on the environment of waste disposal and associated processing mechanisms while reducing further waste output, a new recycling initiative was required.

The existing cardboard extraction and bale production would be intensified to remove a greater volume of cardboard waste by increasing mill-sized bale production and in return raising the level of revenue received from recycling.

Regular producers of bulk food waste from the centre's restaurants and food associated outlets would be targeted to separate food waste in conjunction with other mixed waste material including glass, plastic and aluminium containers. The increase in volume of the recyclable material removed from the general waste stream would enable a reduction in the overall site waste removal costs while reducing the impact on the environment.

PROJECT ACHIEVEMENTS

The additional waste stream separation measures showed a significant increase when benchmarked against the previous year prior to the implementation of the new waste recycling initiative:

- The weight of food waste collected during 2014 showed the total weight removed from site was 32,050 kg. No food waste collection existed until 2014
- The weight of mixed waste collected during 2014 showed a total weight removed from site of 16,380 kg in comparison with 12,705 kg during 2013
- The weight of cardboard waste collected during 2014 showed a total weight of 70,558 kg in comparison with 54,825 kg during 2013
- The number of compacted waste containers consisting of general waste removed from site during 2014 was 17 loads in comparison with 21 loads during 2013. The total annual weight of compacted material removed from the centre during 2014 was 182.94 tonnes in comparison with 215.46 tonnes removed during 2013, showing a total weight reduction of 32.52 tonnes.

WHAT DID THE PROJECT INVOLVE DOING?

In order to achieve significant reductions in the volume of general waste processed and to increase the separation of material to be recycled, business premises producing regular bulk quantities of waste including food, glassware, plastics and aluminium cans were targeted by liaising with store managers to solicit their co-operation and participation in a new waste separation initiative.

Extra separation of waste would be achieved providing additional food waste and mixed waste bins to enable the correct storage and disposal of recyclable material.

The centre`s designated cardboard waste bins would be modified to facilitate ease of placement and removal of cardboard prior to baling. Previously cardboard bin tipping was found to be physically difficult for contract cleaning personnel resulting in less than the total volume contained within the bins being removed on a daily basis.

WHY DID YOU DO IT?

Although waste was being processed with no impact on landfill, the potential for further separation of recyclable material remained and could be achieved by the introduction of new storage containers and waste recovery services.

The initiative would be to reduce waste treatment costs, maximise recycling opportunities and achieve a further reduction in the volume of general waste compacted leading to a decrease in the number of site visits required from the waste disposal provider.

Fewer site visits would lower transportation and disposal costs while benefitting the environment by minimising the number of times a waste vehicle attended site.

WHAT DID IT COST AND WHERE DID THE MONEY COME FROM ?

Funding for the initiative was provided within the allocated waste expenditure contained with the centre`s service charge budget. Revenue from cardboard and savings on reduced load counts would contribute to reducing costs overall.

Cost statistics

Food waste – annual costs during 2014	£2,353
Mixed waste – annual costs during 2014	£2,038
Cardboard bin adaptation (one-off)	£1,625
Cardboard	£3,306
Calculated savings from the reduction in compacted load collection and tonnage charges	£3,505

WHAT DID THE PROJECT ACHIEVE IN TERMS OF SUSTAINABLE DEVELOPMENT, ENVIRONMENT AND/OR EQUITY?

A comparison of figures before and after introduction of the waste initiative clearly show increases in output with the average trend towards sustained growth.

WHO AND WHAT HAS BENEFITED?

The enhanced recycling measures illustrate to the client, managing agent and tenants that green measures and sustainable practices are being implemented, contributing to the centre achieving ISO 14001 Environmental Accreditation in February 2015.

Reductions achieved in waste disposal costs will help to keep waste related service charge costs on or below target

WHAT ARE THE LONG TERM BENEFITS ?

A sustained programme of recycling where bulk quantities can be removed from the general waste stream will demonstrate the centre's compliance with TEEP Regulations reducing the environmental impact of waste generated by the centre for processing.

Continued waste stream separation will enable allocated waste service charge costs to be stabilised for future budget forecasting.

WAS THERE ANYTHING INNOVATIVE ABOUT THE PROJECT?

Involvement in the recycling initiative by both tenants and contact cleaning personnel has enabled them to focus on environmental issues that might otherwise have been ignored and demonstrated through their actions that positive results can be achieved and maintained without major changes to methods of working.

CAN OTHER ORGANISATIONS BENEFIT FROM IMPLEMENTING YOUR METHODS?

The measures undertaken to increase waste stream separation on site can be replicated on a small or large scale by businesses or organisations to address maximising their recycling potential.

WHAT DID YOU LEARN AND ARE YOU PLANNING ANY FURTHER DEVELOPMENT?

Participation by store managers and the dissemination of new procedures to their staff was essential to implement and maintain waste stream separation. In order to maintain momentum on waste recycling measures, the initiative will be expanded to include a greater number of tenants and to investigate further waste stream separation.

SANLAMARE

SEA CHANGE IN BATHROOM WATER USAGE

PROJECT AIM

To drive public knowledge and increase the use of water efficient practices and products in the bathroom at home and at work via the introduction to the UK of the world's most innovative water efficient toilet with integrated hand basin.

The unit recycles and stores water used to wash hands in the basin for toilet flushing at the next use, dropping total water usage in the bathroom by up to 70%. The project is designed to decrease energy used throughout the life cycle of products from manufacturing through shipping, marketing, installing maintenance and disposal, adding to the intrinsic water saving attributes of the product.

PROJECT ACHIEVEMENTS

Since its introduction the Profile5 has gained widespread acceptance in the UK for domestic, commercial, industrial, shipping, sporting, leisure and environmental sectors.

waterwise
AWARD WINNING WATER SAVER

The innovation of integrating a high quality toilet pan, specially designed water recycling cistern, hand basin and stylish tap to provide the world's leading water efficient toilet has gained considerable recognition within the construction industry, both new-build and retrofit, together with local and central governments such as Norwich City Council and the Scottish National Assembly.

As the unit itself provides total water efficiency as well as considerable energy savings, communities have begun to see actual and real savings in both natural resource use and monetary costs.

WHAT DID THE PROJECT INVOLVE DOING?

The project in the UK was ultimately to minimise water use in bathrooms without undermining accepted standards for bathroom products.

Research and development of an innovative water saving technology for the biggest user of water in bathrooms led to the production of the world's first fully integrated toilet suite which includes a hand basin with controlled, measured water delivery, which is then recycled into grey water and stored in the cistern for toilet flushing, resulting in a saving of up to 70% of total water usage in bathrooms. However, the project however had to have mass appeal and offer reliable functionality to ensure general acceptance, together with the lowest possible maintenance costs so as to grow the 70% water saving capacity exponentially.

WHY DID YOU DO IT?

Because we wanted to change the world. We initially set up our company with the sole purpose of identifying and bringing to the UK market world leading energy efficient bathroom products. The project we embarked upon saw us travel the world in search of quality products with measurable, sustainable and replicable natural resource sensitivities, not just in the way they were to be used, but also in the development, manufacture, handling, installation, maintenance and end of life stages.

WHAT DID IT COST AND WHERE DID THE MONEY COME FROM?

To date we have invested almost £1.2m in the UK, which has been completely self-funded. We have recently commenced a feasibility study for the production of additional models/styles of the Profile5 to increase the technology's presence and further positively impact on water and resource use in the UK and European markets. In order to support a rapid growth in demand for the technology, we may, for the first time, have to seek the assistance of grants or investment outside of the company.

WHAT DID THE PROJECT ACHIEVE IN TERMS OF SUSTAINABLE DEVELOPMENT, ECONOMY, ENVIRONMENT AND/OR EQUITY?

Although the project itself has proven successful, with sales now exceeding 1,700 units and with a growth of 60% in the 12 month period from May 2014 to May 2015, we are continuing to promote the use of the product and acceptance of the concept in all UK communities.

This is with a view to having the Profile5 technology, or level of water saving, become a standard to be attained in all new build and retrofit projects. The product is highly sustainable, the water saving attributes it offers are the very definition of environmental sustainability and we believe the economic benefit to individual end users, which is considerable, will be far outweighed by the economic benefit to the UK through total reduced water use and the reduction of carbon emissions throughout the life cycle of products employing this type of technology.

WHO AND WHAT BENEFITED?

Benefits of the project have been experienced by our own company by way of sales, which have ultimately led to the sustainability of the company and our ability to further promote water savings and carbon emission reductions through the development and implementation of desirable products incorporating resource saving technology.

End users have benefited by way of monetary savings in purchase, installation maintenance and water/electricity costs, as well as making a real contribution to the saving of water and other natural resources. Finally, the UK and the global environment have benefited immediately by way of an actual decrease – although we acknowledge of a currently very modest degree – in water use and carbon emissions, as well as arguably the more important benefit of a positively changing community mindset towards water and other resource use.

WHAT ARE THE LONGER TERM BENEFITS?

Continued, exponential growth in water savings as well as significant reductions in carbon emissions arising directly from products employing the technology. In addition, the increase in awareness of the need for and ultimately the actual shift in day-to-day conscious choices for products and processes that have real benefits to the environment and the lives of people globally.

Our goal is to personally help, as well as encourage other organisations, to achieve this readily attainable outcome with the use of stylish and highly functional bathroom products and other homewares that appeal to clients as much for their looks as it does for their environmental credentials.We really do believe that a sea change in saving natural resources has to start at home!

WHAT WAS INNOVATIVE ABOUT THE PROJECT?

The Profile5 was the world's first combination toilet suite and handbasin and it was also the world's first toilet to recycle water used in a handbasin to use for the next flush of the toilet. In addition, the Profile5 utilises the toilet flush buttons to activate the tap which delivers a measured flow of water ensuring that the cistern is only ever filled to requirements, offering even further water savings.

In the first instance every organisation or community that uses the Profile5 toilet suite will benefit from;

- Reduced water usage in flushing toilets at 100% saving
- Reduced water usage in washing hands utilising measured flow
- Reduced carbon emissions in the manufacturing/shipping and handling processes as the unit obviates the need to separately manufacture/ship and handle toilets, sinks and taps
- Reduced carbon emissions in the installation process as the unit obviates the separate costs of installing two separate units instead of the single Profile5 suite
- Reduced natural resource usage in obviating the need to produce, ship, handle and install two sets of incoming water pipes and two sets of outgoing waste pipes in each bathroom scenario the Profile5 suite is utilised
- Reduced maintenance costs
- Reduced end of life costs – not only is the Profile5 100% recyclable, it also saves on decommissioning and removal costs of two separate units as well recycling costs of one unit as opposed to two.

Secondly, other manufacturers can develop similar technology based upon our methods and assist in further reducing water use and other natural resources, together with carbon emissions.

WHAT DID YOU LEARN?

The successful introduction of the Profile5 into the UK has shown us that such technology can be readily accepted by the mass market. The water and natural resource savings were obvious from international testing and implementation, together with the studies conducted during the development process.

However, we have received extensive responses from clients utilising the Profile5 extolling the virtues of and level of water savings, together with lower purchase and installation costs.

We are currently developing other models of integrated toilet suites based on the Profile5 in an effort to increase the technology's appeal in the UK market across all aspects of residential, commercial and industrial applications. ◆

sanlamere

PROFILE™ TOILET SUITE WITH INTEGRATED HAND BASIN BY

caroma
thinking bathrooms

AUSTRALIA'S FIRST 5-STAR WATER-SAVING TOILET SUITE IS NOW AVAILABLE IN THE UK

WATER RATING

QUALIFIED

waterwise

CoRE

sanlamere
aqua-sistemas

AUSTRALIAN WATER-SAVING INNOVATION

No matter which way you look at it, the savings add up with the new Profile™ Toilet Suite with Integrated Hand Basin:

- Water savings of over 70% compared to an 11L single flush toilet suite and separate hand basin
- Water savings of over 17% compared to a standard dual flush toilet suite (6/3L) and separate hand basin
- Water savings of over 10% compared to a Smartflush® toilet suite (4.5/3L) with separate hand basin

Independent field trials have shown the new Profile™ Toilet Suite with Integrated Hand Basin achieves a 10% uplift in water-saving compared to a market leading Smartflush® toilet suite and separate basin. Savings are even greater compared to older flushing technologies.

Water consumption is based on an average home with 2.5 people.

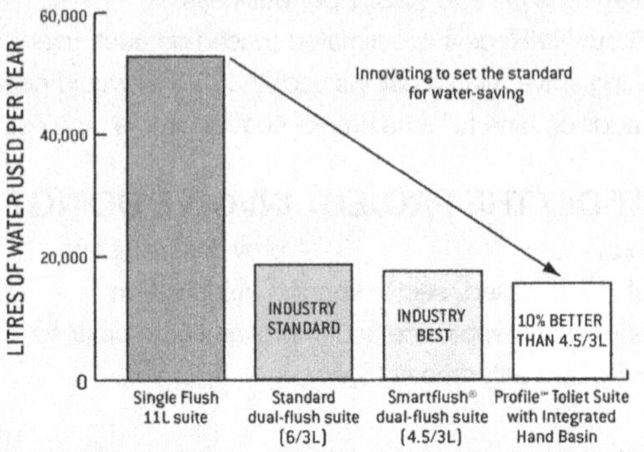

FEATURES AT A GLANCE

- Stylish close coupled toilet suite with Smartflush® technology (4.5/3L)
- Integrated hand basin for enhanced water-saving
- Unique water-saving design
- Space saving design
- Chrome buttons built-in to tapware design
- Available in white vitreous china only
- S and P trap installations for flexibility
- Affordable all in one toilet suite, tap and basin combination

TECHNICAL INFORMATION

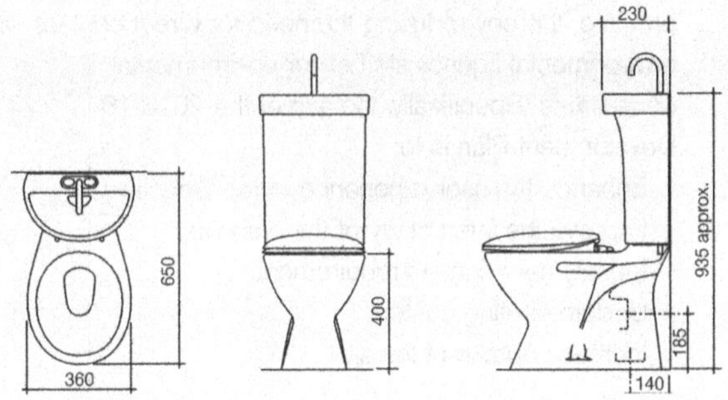

Sanlamere UK Ltd
165 The Broadway
Wimbledon
SW19 1NE
Tel: +44 208 5448 091
Fax: +44 208 1816 893
Email: info@sanlamere.co.uk

SCOTTISH ENVIRONMENT PROTECTION AGENCY

NETREGS

PROJECT AIM

NetRegs aims to improve the compliance and environmental performance of small to medium sized enterprises (SMEs) through the provision of free, sector specific guidance. It aims to present the guidance to its audience in a manner that is easy to access and easy to understand.

NetRegs also aims to continuously improve engagement with business, leading to raised awareness of responsibilities, pro-environmental behaviour change and improved business practices, which will lead to better use of resources and an improved environment.

Through this service NetRegs aims to be a first port of call for SMEs seeking guidance in compliance and best practice, thereby reducing the need for direct contact with environmental agency staff or for environmental consultants. Specifically, the aim of the 2013-16 Development Plan is to:

- Enhance the user experience of the website
- Improve the interactivity of the website
- Identify new content requirement
- Update existing content
- Increase uptake of the site.

PROJECT ACHIEVEMENTS

- Web guidance for 33 business sectors and 12 environmental topics, covering more than 740 different activities
- 15 video clips
- Six interactive E-learning tools
- One compliance self-assessment tool
- On average per month, over the past seven months the NetRegs website received 23,316 web visits, with 56,285 page viewed
- Increase of web sessions year on year since transfer to SEPA in June 2012, with a 77% increase in sessions in March 2015, as compared to March 2013.
- 4,986 subscribers to NetRegs E-update email (since May 2013; a 20 months period)
- 1,727 completed E-Learning courses (Live since February 2014 – a 13 month period)
- 3,300 views of video clips (Live since January 2015- 4 month period)
- 320 self assessments registrations, 819 sections completed (Live since March 2015 – a three week period)
- User surveys show that annual savings to businesses through NetRegs is between £37M and £74M, averaging £1179 to £2231 per business.
- Without NetRegs it is estimated (based on each user making a five minute call instead) that SEPA would need an addition nine full-time employees for simple queries.

WHAT DID THE PROJECT INVOLVE DOING?

NetRegs – www.netregs.org.uk – provides free plain English, web based, sector specific guidance on environmental compliance and improvement to small to medium sized enterprises (SMEs).

It delivers a package of innovative solutions aimed at raising awareness amongst SMEs by using multimedia and multiple channels to deliver powerful environmental messages that will change behaviour and reduce impact on the environment. This is achieved through improving the way guidance is packaged for businesses and extending outreach to a wide business community.

The project was initiated in 2004 as a partnership between the UK environmental regulators. The Transformational

Government agenda and DEFRA policy meant that the Environment Agency (EA) had to withdraw from the partnership and website content and URL ownership was transferred to the Scottish Environment Protection Agency (SEPA).

The SEPA now leads and manages the project and develops and delivers the service in partnership with Northern Ireland Environment Agency. SEPA is applying for this award in recognition for the achievements and developments since the transfer.

SMEs make up the vast majority of businesses in Scotland and are likely to generate more than half the waste and serious pollution incidents, yet the evidence shows they have little understanding of their responsibilities or environmental impact. Regulatory guidance for businesses is often complex and inconsistent resulting in an unnecessary administrative burden, and a lack of clarity about key messages being conveyed. NetRegs was set up to address this.

WHO AND WHAT HAS BENEFITED?

On transferring NetRegs to SEPA, the objectives were to increase the number of users and to develop to ensure specific needs of the Scottish business community were met. To help we set up a Business Advisory Group comprising key business and trade associations as well as some individual businesses.

We also have an Internal Advisory Group to help ensure future policy changes are reflected in our annual guidance review and development plan. Wider engagement is undertaken via regular surveys and meetings.

Since under the management of SEPA , we have built six e-learning tools that deal with common environmental issues. We have also developed a self–assessment tool that allows businesses to do an anonymous individual check to find out if they are compliant. The tools are complimented by 15 videos that give practical examples of actions taken to achieve a high standard of environmental performance. We also send out a free monthly email newsletter to all subscribers.

The 2014 enhancements add to the existing core information provided on the NetRegs website; comprising sector specific information, a library of environmental topics, current and future legislation and links to other useful sources of information. This ensures an easy accessible service providing free, plain English guidance to assist businesses in their day-to-day work.

In addition to maintaining and updating as necessary all existing content we are continuing to develop the site. We are currently developing a mobile app so that users can access information 'off site' and a further 18 videos are under production.

Over the next 12 months we plan to achieve accreditation for our e-learning tools as a formal recognition for continuing personal development (CPD). We will also continue to extend outreach and have begun to implement plans to introduce NetRegs to final year and postgraduate students with the hope they will continue to use NetRegs and introduce it to colleagues when they enter the workplace.

WHAT ARE THE LONGER TERM BENEFITS?

In 2015 NetRegs is transferring to a new content management system which provides an ideal opportunity to improve the look and functionality of the website. A series of focus group covering a variety of sectors will ensure developments continue to be undertaken with customers needs at the forefront of decisions.

Although the geographical spread of (targeted) customers for NetRegs has significantly reduced (with fewer than 10% of businesses being in Scotland and Northern Ireland compared to the whole UK). there has been little change in the volume of guidance. This is because we are still serving the same number of business sectors and there is no reduction in the number of environmental topics. Despite this we have continued to maintain NetRegs with no increase in resource and this includes all the additional work highlighted in this submission.

We regularly monitor the use and value of the website gathering quantitative and qualitative information through analytical data, anonymous and user surveys.

This has provided substantial evidence that the website is being used by a high proportion of the business, the number of users is continuing to grow, users are finding the information helpful in improving compliance and environmental performance and there are also economic benefits associated with its use.

In addition NetRegs contributes to sustainable development through ensuring social inclusion and provides graduate internships on an annual basis.

NetRegs is recognised internally as the main mechanism for providing guidance to SMEs and is included in SEPA's Corporate and Annual Operating Plan. Updates on progress towards key performance indicators are included in Management and Board reports. The small team of staff meet on a regular basis and the enthusiasm and dedication of all involved are key factors in its success.

Although set up for a tool for an external audience, internal staff, particularly reception and call centre staff use NetRegs to answer queries from operators. This brings cost savings for SEPA through diverting calls away from Environmental Protection Officers.

CAN OTHER ORGANISATIONS BENEFIT?

The NetRegs model is applicable to any country and could be easily replicated as an information based tool for improving environmental performance. It has been named as a good practice example by many organisations across Europe including OECD, European Commission and IEMA. The team regularly provides advice to other organisations on the NetRegs approach to providing guidance. Examples include presentations delivered in Ukraine and Moldova earlier this year.

Key achievements for NetRegs include 23,316 unique visits per month, equating to a 77% increase in the use of the website since transfer to SEPA two years ago. User surveys show that annual savings to businesses through using NetRegs is between £37m and £74m, averaging £1,179 to £2,231 per business.

NetRegs

For Northern Ireland and Scotland
www.netregs.org.uk

Guiding small businesses through environmental regulations

Understand your environmental regulations and learn about best practice

NetRegs provides free practical guidance for your business with over 30 guides for specific business sectors, such as agriculture, construction, chemicals manufacturing, retail, offices, etc.

- Information is provided on topics such as emergency planning, air and noise, storage and handling of materials, packaging, energy efficiency, renewable energy options, waste management and the protection of water.
- You can sign up to receive monthly email updates from NetRegs, keeping you up to date with new regulations and other news items that are of interest.
- E-learning tools, self-assessment tool and best practice video clips now available.

SEPA
Scottish Environment Protection Agency

NIEA Northern Ireland Environment Agency

www.netregs.org.uk

SKANSKA

NORTH WEST CAMBRIDGE DEVELOPMENT

PROJECT AIM

The North West Cambridge Development is a large mixed-use development on the outskirts of the city of Cambridge. The vision is to create a new district and extension to the city, centred on a mixed academic and urban community. Located two miles to the north west of the city on a 150 hectare site, development works have been split into several phases. The first phase will see the delivery of housing and community facilities. Designed by Aecom for the University of Cambridge, Phase 1 is being constructed by Skanska as Principal Contractor for the Site Wide Infrastructure.

PROJECT ACHIEVEMENTS

The site includes the largest rainwater recycling system in the country that aims to minimise the risk of localised flooding in an area that is already prone to flooding, while also reusing the water to reduce potable water consumption per person across the whole development.

One of the largest underground bin systems in the UK will be in place in support of the site-wide sustainability aspirations. A traditional system on such a development might see 9,000 wheelie bins being emptied on a regular basis. The solution is 450 underground bins which will service the waste produced by the 3000 homes and reduce carbon by taking away the stop-start collection of traditional systems.

It was identified early on that disruption to the local community must be minimised as far as possible and the construction should blend with the local landscape. To meet this 350,000m3 of soil has been retained on site via cut and fill construction.

The site has been constructed to mitigate the impact on wildlife during construction as well as improve habitat once complete. Newts, water voles and kingfishers are just some of the species that will benefit.

SKANSKA

WHAT DID THE PROJECT INVOLVE DOING?

The North West Cambridge Development is a large mixed-use development on the outskirts of Cambridge. The vision is to create a new district and extension to the city, centred on a mixed academic and urban community.

Located two miles to the north west of the city on a 150 hectare site, development works have been split into several phases. The first phase of the development will see the delivery of housing and community facilities. Designed by Aecom for the University of Cambridge, Phase 1 is being constructed by Skanska as Principal Contractor for the Site Wide Infrastructure.

This involves construction of roads and transport routes, foul and surface water drainage, sports facilities as well as public, green space and landscaping, incorporating the largest water recycling system in the UK.

Also working on the site are seven other principal contractors responsible for the building lots, including accommodation, primary school, supermarket, hotel, retail units and a community centre. Phase 1 is due for completion in spring 2017 with further phases to follow.

WHY DID YOU DO IT?

The University of Cambridge is one of the world's leading universities. It is renowned for the excellence of its teaching and research, and it makes a significant contribution to the prosperity of the city of Cambridge and the wider UK economy. To maintain its reputation the university must continue to develop and grow. In particular the university needs to address thelack of affordable accommodation for staff and postgraduate students and the need to foster research and development links.

WHAT DID IT COST AND WHERE DID THE MONEY COME FROM?

The overall development is projected to cost in the region of £1billion with Phase 1 estimated at £281million. The project is being funded by the University of Cambridge who also own the site.

WHAT DID THE PROJECT ACHIEVE IN TERMS OF SUSTAINABLE DEVELOPMENT, ECONOMY, ENVIRONMENT AND/OR EQUITY?

In creating the new site, there is an opportunity to affect the built environment and how people lead more sustainable lives. Innovative measures that are unique to this development include the largest water recycling system in the UK, a district heating network and an underground bin system. Homes and non-residential buildings will be built to some of the highest levels of sustainability.

It was identified early on that disruption to the local community must be minimised as far as possible and the construction should blend with the local landscape. To meet this 350,000m3 of soil has been retained on site via cut and fill construction.

The site has been constructed to mitigate the impact on wildlife during construction as well as improve habitat once complete. A unique specially constructed newt tunnel was installed underneath one of the entry routes so as to enable great crested newts to access habitat and breeding ponds either side of the road.

Artificial kingfisher nest boxes are to be installed in the banks of a watercourse to increase numbers and the river bed will be constructed so as to provide suitable habitat for water voles.

In addition features to encourage the use of new buildings by roosting bats have been incorporated into the design of the sports pavilion and utility buildings. A third of the development (50 hectares) will be open space for residents and users to enjoy. This will be achieved by planting more than 2400 trees by the end of Phase 1.

WHO AND WHAT BENEFITED?

All users and residents of the development will benefit from the commitment to creating a sustainable legacy. The primary school and its related infrastructure will be opened ahead of the rest of the development in time for the 2015 academic year and prioritising housing will aid the housing supply shortage in Cambridgeshire. The development includes extensive cycle and pedestrian routes to promote connectivity around the site itself and link with the city centre.

WHAT ARE THE LONGER TERM BENEFITS?

The development will promote education through being a living demonstration of a sustainable community, or a 'living experiment' for ongoing research and monitoring. New and innovative technologies and systems will be on-show for residents and visitors to explore, supported by easily accessible information and open days.

For example, energy and drainage infrastructure will be displayed in some areas for people to view and the geological SSSI that lies on the eastern boundary will be incorporated into the site so users can see and learn about this special area.

The green space will create a lasting legacy for the local community to enjoy and the investment in drainage will lessen any future impact of severe weather and climate change.

WAS THERE ANYTHING INNOVATIVE ABOUT THE PROJECT?

The site includes the largest rainwater recycling system in the country that aims to minimise the risk of localised flooding in an area that is already prone to flooding, whilst also reusing the water to reduce potable water consumption per person across the whole development.

Making use of SUDS, hard and soft landscaping areas store and slowly release water through the development. Rainwater is then stored in specially designed 'green landscape' areas and treated before it is pumped back into the homes for non-potable water uses, such as toilet flushing and irrigation. Furthermore, large areas of land will be purposely designed as flood plains in the eventuality of a one in a hundred years storm.

Skanska also utilised temporary construction SUDS to maximise the retention of surface water run-off. A series of interconnected swales and balancing ponds were used to collect and hold surface water over winter. This was subsequently then used for dust suppression and watering of plants over summer. During the first five months of 2015 alone, more than 1.4million litres of collected surface water (non-potable) has been used rather than drawing from the mains (potable water).

One of the largest underground bin systems in the UK will also be in place in support of the site-wide sustainability aspirations. A traditional system on such a development might see 9,000 wheelie bins being emptied on a regular basis. The solution is 450 underground bins which will service the waste produced by the 3000 homes and reduce carbon by taking away the stop-start collection of traditional systems.

CAN OTHER ORGANISATIONS BENEFIT FROM IMPLEMENTING YOUR METHODS?

Knowledge and best practice from all principal contractors on site has been shared with the university during monthly progress meetings to raise standards and improve performance. The lessons learnt will develop and feed into future Phases of the development.

WHAT DID YOU LEARN AND ARE YOU PLANNING ANY FURTHER DEVELOPMENT?

The university is seeking approval to proceed with a feasibility study for Phase 2 which will be constructed using the same ethos and sustainable principles.

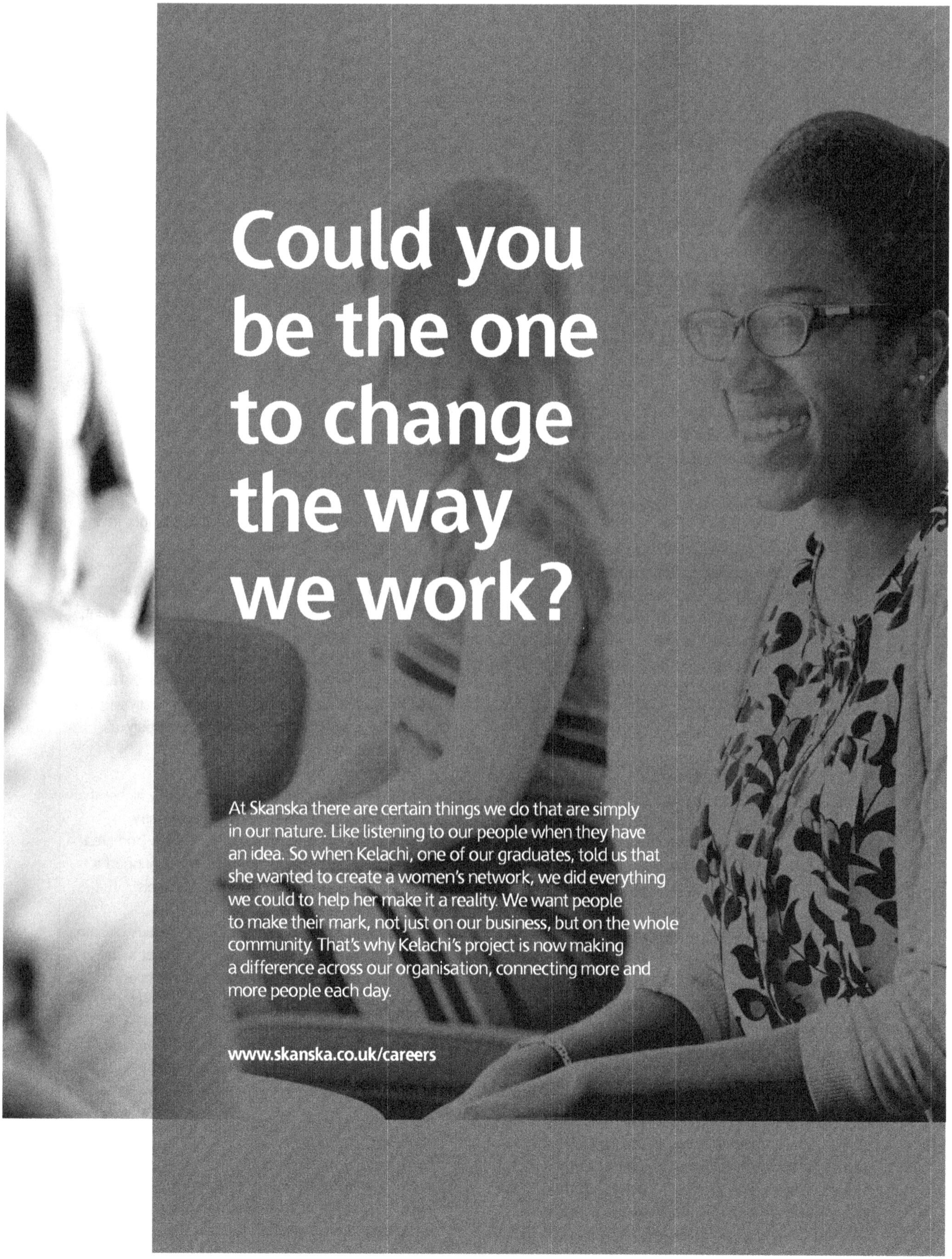
Could you be the one to change the way we work?
At Skanska there are certain things we do that are simply in our nature. Like listening to our people when they have an idea. So when Kelachi, one of our graduates, told us that she wanted to create a women's network, we did everything we could to help her make it a reality. We want people to make their mark, not just on our business, but on the whole community. That's why Kelachi's project is now making a difference across our organisation, connecting more and more people each day.
www.skanska.co.uk/careers

SKANSKA

SONNE ENERGETICOS, MEXICO

CARBON FREE COLLECTIONS

PROJECT AIM

Working in partnership with the Green Earth Appeal, Carbon Free Collections delivers a comprehensive and verified carbon offsetting solution for their collection vehicles which visit their customers, including KFC McDonalds and Walmart.

With 15 years of experience, Sonne Energeticos provides a used cooking oil (UCO) collection service across Mexico. We are fully approved by several government and environmental regulations that are necessary for collecting, transporting and safely processing the grease waste/used cooking oil from restaurants and other food service industries. All fat/cooking oil collected is recycled into new products that are environmentally friendly.

PROJECT ACHIEVEMENTS

All the used oil collected is processed and recycled for its reuse in different industries such as the chemical industry or for the production of biofuels, in the most safe, responsible and efficient way that reduces environmental impact and promotes sustainable development.

In relation to these business activities Sonne Energeticos aims to turn their negative carbon impact into a positive by planting trees in conjunction with the Green Earth Appeal. By offsetting their carbon footprint Sonne Energeticos effectively enables their customers to have a positive carbon impact on the environment.

WHO AND WHAT HAS BENEFITED?

Sonne Energeticos operates the most clean and technologically advanced oil collection fleet in the market, clearly identified with the company name and logo and driven by uniformed operators. Many of their vehicles run on biodiesel fuel, produced from the used cooking oil collected. This way, a recycling model is created, which is the essence of sustainable development.

The first two major beneficiaries are to the planet and the third is to some of the poorest communities in the developing world, which covers the cost of planting a single tree in Africa, Asia or South America. The first benefit to the planet is reforesting areas which have become barren and where desertification has begun to take hold. Planting quick growing, hardy Moringa trees promotes the water table, thereby enabling further fruit trees to be planted.

The second benefit to the planet is that each of the trees planted will absorb carbon dioxide from the environment, reducing the negative carbon impact associated with the transportation of goods and services.The third benefit aids the poorest communities in the developing world. The organisations who facilitate the tree planting firstly work with local farmers, educating them on Agroforesty, a method of achieving maximum yields.

Biodiversity in Agroforestry systems is typically higher than in conventional agricultural systems. Agroforestry incorporates several plant species into a given land area and creates a more complex habitat that can support a wider variety of birds, insects, and other animals.

Agroforestry also has the potential to help reduce climate change since trees take up and store carbon at a faster rate than crop plants. Once the farmers have been through workshops, materials are provided to set up tree nurseries. These nurseries negate the need to ship in trees, emitting large amounts of carbon in transportation. It also enables the communities to take full ownership. Once the nurseries are working to capacity, they are able to grow tens of thousands of saplings ready for planting in nearby Agroforestry areas.

Thanks to Sonne Energeticos, communities in the developing world are empowered rather than left fighting for sacks of aid from the back of trucks. The planet benefits from high volumes of trees being planted with all associated benefits.

SOUTHWAY HOUSING TRUST

GREENING OUR GREY CITY

PROJECT AIM

- To protect and preserve our urban green spaces from being converted to grey while adapting to our urban lifestyles of increased vehicle numbers and people's desires to park their cars in close proximity to their properties and off the carriageway
- To support access across green spaces that we own to private home owners with no impact on our business connected to costs and ongoing maintenance and liability. while retaining the green infrastructure and ownership of the land
- To support the MIAP application process for access to driveways without the loss of green space
- To be able to offer a viable SUDS scheme and help towards reduction of storm water run-off, work towards reducing the heat island effect of our urban environments and ensure that our urban environment remain aesthetically pleasing to ensure the area is a desirable place to rent.

PROJECT ACHIEVEMENTS

- Carried out a successful partnership project with MIAP to provide a drive within curtilage of garden space and a grass reinforced mesh access driveway across our green space in front of the property, enabling access for a tenant with mobility restrictions
- Converted an unsightly part grass, part concrete and partly mud area on a main carriageway into Manchester city centre into an area that now offers a SUDS scheme and ensures the route into our city is aesthetically pleasing whilst converting hard standing back to green space
- Written a strategy and policy that enables private home owners to work alongside us as a land owner to access their properties that has previously not been an option. Ensured that there is no long term maintenance costs to the business or litigation implications
- Protects the aesthetics of our neighbourhoods to ensure that the value of our housing stock is enhanced to its full capability
- Ensures the protection of green spaces for people that live within our neighbourhoods to reap the benefits connected to health and well being
- Enables us to carry out enforcement on tenants as well as private home owners who are damaging our land if engagement is not wanted by the perpetrator through a phased approached with Greater Manchester Police and our legal advisors if need be.

WHAT DID THE PROJECT INVOLVE DOING?

Although using reinforced grass mesh is not a new technology, using it as a material for the purpose of drives across grass verges or creating areas for parking on which once would have been grass or is presently grass is not common practice.

Councils and housing providers across the UK all suffer with problems connected to parked cars, damage to grass spaces, silt run-off and then localised flooding through blocked drains and then the impacts connected to degradation of tarmac with water ingress.

As this is not common practice, a lot of time was spent on our estate understanding this issue at every level, and a lot of research and partnership working was needed with both Greater Manchester Police and our legal advisors to write the policy and procedure in the first instance.

Internally we have needed to work with a variety of teams across the organisation to provide information and training in the delivery of the policy but also around a need for a cultural change with in the organisation as a whole.

With this environmental management situation being quite complex and messy, it has taken a lot of thought regarding the policy and the implementation of this policy to manage community expectations as well as the differing views of people that want to park on the grass and are happy to churn it up and people that love and care for their green spaces.

We have spent a lot of time talking to perpetrators to explain the situation as a whole and point out that knock on effects like localised flooding which they don't like and on most occasions had not connected the two activities together. The main element of this process has been about trying to find a workable option to our changed environment and needs but championing our green spaces in our ever greying city.

WHY DID YOU DO IT?

There where three main drivers behind this project. The issue first came to light as we were setting up our health and safety inspection regime. There were a variety of elements connected to this:

- The trips and risks associated with dried out hard ruts in the summer and slippery muddy areas on our green spaces and protection from claims re repairing these areas
- This is not a sustainable process due to the fact that the root of the problem has not been addressed so an appropriate solution to the problem was needed
- Looking at present common practice of using tarmac, this opened up other concerns regarding SUDS, heat island effect and contributing to greying our greenery.

The second driver was the wider environmental impacts that this behaviour was creating. This varied from localised flooding, erosion, damage to tree roots that were protected by TPOs, the aesthetics of the area being damaged and the impacts connected to health and wellbeing associated with green spaces being reduced as we turn them into car parking areas or drives.

The third driver was connected to community cohesion. The situation in some areas was quite complex and messy when it came to some people wanting to drive on the grass and park outside their homes and other residents who were dead against this behaviour as well as despised the impacts that this behaviour was creating

As the areas became worse over the winter periods, it also encouraged others to drive over the areas as the area all ready looked in a poor state of repair – "So one more person won't make a difference".

WHAT DID IT COST AND WHERE DID THE MONEY COME FROM?

If we are working along side MIAP, the funding for driveways if a tenant or home owner meets their criteria comes through them as an organisation but this funding is already in place to assist with access requirements. We have just changed the spec of materials used.

If we are working with a private home owner, the funds to deliver the work needed, including our staff time comes from them. If we are looking at installing drives for our tenants, the funding is provided through the rent that is paid to us as a Housing Association.

The staff time to set this project up has been funded through Southway Housing Trust and driver for this has been increasing insurance claims and pay outs as a business. These have significantly reduced over the last 18 months since our inspection process and repair process has been implemented.

WHAT DID THE PROJECT ACHIEVE IN TERMS OF SUSTAINABLE DEVELOPMENT, ECONOMY, ENVIRONMENT AND/OR EQUITY?

The project has enabled both tenants and home owners to access this scheme where previously as a home owner there was not an option for any driveway to be installed on our land due to issues of maintenance and litigation.

The scheme as a whole falls under Sustainable Urban Drainage which is deemed as good practice and works towards sustainable management of our urban estates, our environment and supports the adaptations needed for climate change.

Investment in green spaces has been shown to help improve a region's image and will help attract and retain high value business, new business start-ups and workers. This in turn increases the scope private sector investment, reducing unemployment and increasing GVA as well as land value and property value.

WHO AND WHAT HAS BENEFITED?

Neighbourhoods that fall within South Manchester in the areas that our land sits are the locations where this policy has been implemented and has benefited from. Within these areas we own 157 green spaces so all properties that surround these areas have the potential to be able to benefit from this scheme either directly or indirectly.

The benefits are far reaching from safer spaces, well managed estates that then enable us to provide educational, exercise sessions and fun activities on our green spaces with local schools and for members of the local community.

These measures should reduce isolation, increase people's mental health and fitness levels, and improve community cohesion. In turn, Manchester as a city has communities that have a better quality of life and a positive experience living.

WHAT ARE THE LONGER TERM BENEFITS?

For the business long term, our insurance claims will continue to decline which in turn will reduce the premiums that we need to pay. In 2013/14 we had five reports of claims, but information was sketchy and we were not able to gather information to defend the claim.

In 2014/15 we have had 17 reports of potential insurance claims, eight of which we have been supported by full claims which will ensure that if we do get a claim that we will be in a very good position to defend ourselves.

WAS THERE ANYTHING INNOVATIVE ABOUT THE PROJECT?

Although the material of grass reinforced mess is not a new product to the environment, the process of using this to combat parking issues and life style choices is a new process.

Councils and housing providers preferred options are to tarmac to create parking areas or drives or to simply leave the area and do nothing. We however are actively engaging with the situation to provide options and ensure that the pressures on urban land use are well designed and well managed.

CAN OTHER ORGANISATIONS BENEFIT FROM IMPLEMENTING YOUR METHODS?

As an organisation we have had a very positive benefit financially through our insurance. Our insurance company has reported that in the last 12 months we have made really important progress in being able to effectively defend claims.

Of the 17 current personal injury claims we have live in the system, eight have been denied and we have been able to provide excellent supporting evidence to defend ourselves. We have admitted liability on two and we are in the process of investigating and denying liability on the remaining seven.

This is a massive improvement on the position we were in 12 months again where, due to lack of formal reporting and clear environmental management procedures, we were admitting liability on claims that we should have been able to defend.

A positive by-product of this is that people are suffering fewer injuries and there is a reduction in accessing services such as the NHS or people taking sick leave. The behavioural trait being demonstrated of driving and parking on grassed areas is not unique to this area so could be adopted in any urban area.

WHAT DID YOU LEARN AND ARE YOU PLANNING ANY FURTHER DEVELOPMENT?

We learnt that people are generally very open to the options available to them and that there is quite a large percentage of people that drive or park on grassed areas because their neighbours do.

Once informed that this is no longer an option, people have mostly started to park on surrounding roads. There has also been a link to people then not parking on footpaths, which means pedestrians with push chairs do not have to walk in the road to get round the cars.

We are working to roll this out right across our area on a case by case basis to ensure that we can deal with each situation on its own merit until we are in a position to move onto the next problem area. We have found this is the best way of dealing with each case to ensure the time is given to work through the problems and issues in a quick and efficient process.
For homeowners that want to the option of installing an access route to their property, we have found that they are grateful for the technical support and the project management side of the work being completed.

With the work that we have carried out in the research phase with GMP as well as with our legal team, we have ensured that we have covered our legal rights and responsibilities towards managing this situation as well as highlighting what risks are present re the 20 year right of access.

This would still have to be tested as we have not come across this to date as part of this process.

A TALE OF TWO CITIES: LIVERPOOL & MANCHESTER

We won the Grow Wild England flagship Award in 2014.

Southway Housing Trust is proud to be part of this winning Grow Wild England Landmark Project, which has improved landscapes in both cities through the use of stunning wildflowers.

This project continues to bring people and communities together in both cities to sew, learn about and experience the beauty of wildflowers.

We are proud to be partners in making a difference through improving nature and landscapes around us.

'2016 holds unlimited promise with more diverse species and more people coming aboard to tell a new Tale of Two Cities!'

Find out more and visit:
https://www.growwilduk.com/community-projects
http://www.nwc.org.uk/landlife/about_a_tale_of_two_cities

SUBANG JAYA MUNICIPAL COUNCIL, MALAYSIA

COMMUNITY EDIBLE GREENS

PROJECT AIM

To collaborate and engage our communities in our green mission so they do not just see trees, shade and oxygen as the only incentives.

Our communities need additional tangibles such as food, organic vegetables, fruits and herbs.

We came out with a simple blueprint for edible greens. It is modulated and regulated project with multiple bottom line results to achieve:

- Efficient greening of land spaces within dwelling areas
- Attractive and rewarding green through planting of edible plants
- Platform for community collaboration and engagement
- Promoting economic and social gains.

We aim at two community groups:

Landed property residents

Given the tropical nature of our environment, easily about 160,000 landed property residents maintain simple or elaborate gardens with assortment of edible and ornamental greens. There are also those with unsightly and neglected gardens. The latter becomes our priority focus to curb health threat and constant complaints from other neighbours.

The future is green: landed property residents maintain simple or lavish gardens under the initiative

Green shoots: residents of Subang Jaya have a better diet thanks to growing vegetables

Flat and apartment dwellers

Our project feasibility survey revealed that another great source of genuine interests in green activity is dwellers of low cost flats or apartments. They have no land to manifest their green passion. Their engagement interests in green activity are brimming with interests and in abundant supply.

PROJECT ACHIEVEMENTS

After almost two years, the result is work in progress. We devised a simple Green Neighbourhood Planning Guidelines which make our edible greens project modulated and regulated:

Landed property residents

- Special recognition for residents with well-maintained greens within their premises
- Free advice and other incentives for those planting edible greens
- Initial soft approach by offering advice and assistance for those with neglected gardens. Helping them switch to edible greens creates much longer enthusiasm due to the economic value.

Flat and apartment dwellers

We have gathered 360 families tending 360 plots of edible greens. The number is growing at 20 to 30 plots a month. The project serves as a prototype that has shown tangible results. These include a measurable increase in greenery within the flat and apartment common areas, and potential increment of greenery coverage within and around private dwelling places.

The plot owners have achieved 100% sufficiency in vegetable requirements and could even share their surplus with others. It also contributes to lesser traffic congestion and lower carbon emission. The projects serve as a mindset shift in favour for greenery and create substantial demand for compost from our biomass centres.

WHAT DID THE PROJECT INVOLVE DOING?

It is a green collaboration through community engagement in planting edible greens within their dwelling areas. The project is modulated and regulated based on our Green Neighbourhood Planning Guidelines.

The project focuses on two community groups. One is for 160,000 households living in landed properties spread around our 168 sq km boundaries. The other project is for 33,619 households of low cost flats and apartments.

Edible Greens Project for Landed property residents entails the following activities

- Special recognition for residents with well maintained greens within their premises
- Free advice and other incentives for those planting edible greens
- Initial soft approach offering advice and assistance for those with neglected gardens. Helping them switch to edible greens generates enthusiasm due to the economic value.

Edible Greens Project for flat and apartment dwellers

With the 360 families tending 360 plots as on goings, we shall approve and sponsor new applications at the rate of 20 to 30 plots a month.

Applications are through resident associations to use vacant spaces within or nearby their dwelling areas. They only need to form a family group of minimum six or seven committed and garden loving passion to qualify for a plot each. Participating families also pledge to plant and maintain the garden as a group effort. They are given the choice to plant any edible plants or vegetables on their plots as long as the harvests are geared for home consumption only and their plots are regularly cultivated for continuous greenery and their gardens maintained to be aesthetically attractive.

Our landscape officers make regular rounds to motivate and to lend support. We establish strong collaboration with local agriculture agency who readily gives expert advice and free supply of seedlings.

WHY DID YOU DO IT?

We want to be a fully fledged Green City by 2030 with multiple bottom line results to achieve. Our success has to depend heavily on the quality of our green collaboration with the community we serve.

Our random feasibility survey shows grassroot communities want tangibles like food, organic vegetables, fruits and herbs among their greens too. We decided the commonality lies in the edible green projects.

While we seriously pursue community engagement in our green efforts, issues like the misuse of ground space, threats to public health and unsightly and neglected gardens have to be managed according to the formulation of our Green Neighbourhood Planning Guidelines.

WHAT DID IT COST AND WHERE DID THE MONEY COME FROM?

The council sponsors only the initiation of the project which involves approving land spaces, plots and other basics. Our Landscape Department starts with a budget to sponsor a three times three metre wooden box filled with fertilised soils, rainwater harvesting containers, compost bins, common fencing and seedlings. We have spent $30,000 to provide plots for the first group of 360 families.

For the landed property households our service is mainly providing green advice and supplying edible green seedlings. Our biomass centres provide compost.

WHAT DID THE PROJECT ACHIEVE IN TERMS OF SUSTAINABLE DEVELOPMENT, ECONOMY, ENVIRONMENT AND EQUITY?

- Our measurable increase in greenery shall include public areas and increment of quality greenery coverage contributed by private dwelling places
- The edible green project which started earliest have already enjoyed several harvestings. The plot owners have achieved 100% sufficiency in vegetable requirements and could share their surplus with others
- Another concrete step to lessen traffic congestion and lower carbon emission when residents have one less reason to make regular trips to buy vegetables
- The projects serve as a mindset shift in favour for greenery. It serves as educational platform to educate the grass root community, especially the young. Using this platform we shall gain other form of collaborations which are vital for our success
- The participating residents also shared their satisfaction and delight not only due to fresh supply of organic vegetables, but the social harmony the project creates.

WHO AND WHAT HAS BENEFITED?

The project beneficiaries are as follow:

- The Council has found a novel way to achieve its green mission to be the Green City by 2030
- The community benefits economically and socially. The project also enhances the neighbourhood bonding among the community
- The traffic congestion problem is provided with another potential solution.

We foresee a longer term benefit in health provided by the project's organic vegetables replacing the market supply of 'pesticide vegetables' that have untold health damage and healthcare costs. Better quality collaboration will pave the way for more community engagement and the desired effect of having a more civic minded society.

WAS THERE ANYTHING INNOVATIVE ABOUT THE PROJECT?

We believe the innovation lies in the green collaboration which creates community engagement needed for the success of our green mission. It's a win-win situation for the council and the communities.

CAN OTHER ORGANISATION BENEFIT FROM IMPLEMENTING YOUR METHODS?

Yes. We hope to set an example of green collaboration for others to follow. More of such projects will benefit the country. Malaysia is a net importer of vegetables despite our tropical blessings. Our mass cultivations are dedicated largely to oil palm. We believe this is the answer to self sufficiency in edible greens consumption.

We are very proud that this project has already been implemented in schools. We look forward to help creating our new green generations.

WHAT DO YOU LEARN FROM THE PROJECT AND ARE YOU PLANNING ANY OTHER DEVELOPMENT?

The success of our project hinges on our ability to modulate and regulate within the communities we operate. We hope our success breeds more success. We plan to incorporate better technology in soil management, planting, maintenance, and harvesting.

TEAM VAN OORD

MEDMERRY MANAGED REALIGNMENT

PROJECT AIM

The aim of the Medmerry scheme was to provide sustainable flood risk management, taking into account predicted sea level rise, whilst maximising the opportunity for habitat gain and engaging the community. The scheme integrated the design of habitats into the design of the flood risk management allowing multiple benefits to be achieved in a sustainable way.

In providing improved flood risk protection and creating an accessible nature reserve, the project also aimed to provide economic benefits to the area.

Holding back the tide: Medmerry scheme provides flood risk management

PROJECT ACHIEVEMENTS

Economic

- Protection of 348 properties, sewage works and local caravan parks on which Selsey's economy is dependent
- Protection of Selsey's main road serving more than 5,000 residents
- £300,000 per annum saved on flood defence maintenance
- £90m direct economic benefits
- Local caravan parks have extended opening periods by two months, generating income and jobs for the area.

Environmental

- 183ha intertidal habitat, ensuring resilience for other losses in the Solent
- 300ha habitats of principle importance
- 1,000 reptiles were moved to safe new habitat
- Resilience protecting existing and new habitats – long term sustainability
- Formation of a fish nursery, to sustain the local fishing fleet.

Social

- Improved public health associated with access to countryside and wildlife
- Giving people the opportunity to influence the creation and development
- 10km of new access routes
- Opportunities for green tourism
- Improved knowledge of the history of site through archaeological excavation.

Sustainable construction

- 3D model of site produced linked to detailed materials management plan
- Real time GPS data on plant optimised excavation activities

- 100% excavated material reused on site
- 60,000t rock delivered by sea, minimising impact on the local road network
- 20,000t of recycled aggregate used.

WHAT DID THE PROJECT INVOLVE DOING?

Medmerry has long faced problems with flooding from the sea, with regular breaches of the shingle bank, most recently in 2008 when more than £5m of damage was caused.

Medmerry delivers reduced flood risk, compensatory habitat and long term community benefit and is one of the most sustainable projects the Environment Agency has delivered.

Seven kilometres of new embankment were built and the existing defences breached to create a new wetland habitat. The scheme provides flood protection to 348 properties, sewage works and the main road which serves more than 5,000 residents.

Medmerry provides a major boost for green tourism and is managed by the RSPB as a nature reserve, providing access, education and volunteering opportunities at the same time as securing long term environmental benefit for the area.

WHO AND WHAT HAS BENEFITED?

As a result of rises in sea levels, internationally important coastal wetland habitat is being lost across the Solent. Medmerry is one of very few sites with the potential to create an intertidal habitat on a large scale and one that has been designed to be resilient to sea level rise for at least 100 years. Medmerry provides 183ha of intertidal habitat, enough to meet European targets for the next 20 years. To safeguard protected wildlife on site like water voles and reptiles, we created 300ha of habitats of principle importance.

WHAT WAS INNOVATIVE ABOUT THE PROJECT?

Medmerry is the UK's first managed realignment scheme on open coast. This groundbreaking project has been recognised locally, nationally and internationally as an exemplar project of innovation.

In addition to the innovative approach to managing coastal flood risk, this project has developed innovative approaches to delivering ecological mitigation including:

- The RSPB worked with tenant farmers to create a "push pull" cropping regime. "Push" crops, those unattractive to birds, were planted in construction areas, encouraging birds to move to "pull" areas outside the construction area where crops attractive to birds were planted. This successful strategy ensured no construction delays due to ground-nesting birds. During the 2013 season, lapwing, avocet, ringed plover and little ringed plover all successfully fledged chicks
- A conservation strategy was licensed for managing the scheme's construction and operational impacts on water voles. Extensive new freshwater habitats were created landward of the new flood embankments. The strategy for water voles used natural processes and materials and was integrated with the flood risk management design process to ensure long term conservation of water voles at Medmerry.

WHAT ARE THE LONGER TERM BENEFITS?

Planning Medmerry began in 1998. Environment Agency staff required a big vision and huge commitment to deliver the scheme. The same staff involved in the planning have seen the scheme through construction and into the operation and monitoring phase. Medmerry could not have been delivered without the enthusiasm and commitment of senior staff and the support of the local community.

Social benefit

The project team worked closely with local people to design more than 10km of new footpaths, cycle paths and equestrian routes in an area that previously had little access. Carefully positioned viewpoints around the site will ensure visitors gain maximum enjoyment from the site without causing any disturbance. Interpretation at these viewpoints will help visitors understand what they are seeing, and the history of Medmerry. Local access groups were engaged to ensure that the new routes are inclusive.

Economic benefit

The scheme was delivered for £20m, with direct economic benefits of £90m. Eliminating the need for intensive ongoing maintenance saves approximately £300,000 per annum. The project was funded by central government Flood and Coastal Erosion Risk Management Grant in Aid.

The completion of Medmerry is a major boost for green tourism. As a result of the enhanced flood protection, the large local caravan parks can now extend their opening periods by two months, generating income and job security for the area.

As the wetland habitats develop over the years it is likely that people will visit these caravan parks specifically to watch the wildlife, opening up potential new tourist markets locally.

Recent extreme weather events serve as a reminder of our changing climate. Medmerry defences stood up to the worst storms for 20 years, fully protecting the areas that would previously have faced extensive flooding.

RSPB monitoring shows that birds are already using the new habitat extensively. In the summer of 2014 a pair of Black-winged Stilts successfully fledged three chicks on the site. This is the first time since 1987 that they have successfully bred in the UK.

ARE YOU PLANNING ANY FURTHER DEVELOPMENTS?

As part of the community discussions about the access at Medmerry, there were many aspirations that the scheme could not deliver.

These included extending the new access routes away from Medmerry to link with other established tourist destinations and tourist routes, as well as encouraging new initiatives to increase green tourism to the area.

A community and local council initiative has formed the Green Links Around Medmerry (GLAM) partnership. The core objectives of this group are to expand upon the access infrastructure at Medmerry to create a wider access network, link the communities in the area and create a sense of place and ownership for local residents.

Access to the countryside and being amongst wildlife has been shown to bring public health opportunities and benefits. Under the management of the RSPB and in partnership with GLAM and MStAG, these opportunities will continue to be explored and expanded at Medmerry, and will be one of the project's major legacies.

CAN OTHER ORGANISATIONS ADOPT YOUR POLICIES?

Rising sea levels and the economic challenges of maintaining coastal sea defences will make coastal realignment an increasingly sustainable option in the future.

The uniqueness, scale and innovative ecological approaches implemented on this scheme are influencing Environment Agency policies on the economic and environmental benefits of managing coastal flood risk through managed realignment. Innovative ecological mitigation techniques developed during this project have been documented and can therefore inform the design of future managed realignment schemes, reducing cost and increasing the chances of success.

Delegations from Japan, Malaysia and the Netherlands have already visited Medmerry to learn first hand from our experiences.

THE CRYSTAL

THE CRYSTAL

PROJECT AIM

The Crystal is an initiative by Siemens, purposely built in 2012 to demonstrate the important role that technology can play in creating a sustainable future. It is located in London's Royal Docks, a location with huge sustainable potential, and aspirations from local stakeholders to improve the sustainable footprint of the area.

Shining example: The Crystal acts as an inspiration and educational tool

As well as supporting the regeneration of East London, the venue also acts as an inspiration and educational tool for visitors and local communities; highlighting that The Crystal is evidence that while sustainability is a challenge, the solutions to achieving it are within our grasp.

PROJECT ACHIEVEMENTS

It is the first building in the world to gain the highest ratings in both LEED (Platinum) and BREEAM (Outstanding), the world's two leading accreditation bodies for sustainable buildings. On top of building accreditations, The Crystal has series of certifications which ensure its operational management upholds these high standards and meets the requirements of the customer, such as the international standard ISO 20121, ISO 14001 and OHSAS.

On top of this, we boast a Gold Green Tourism Award, RICS (Royal Institute of Chartered Surveyors) Award, BCO for excellence in office spaces, LABC London Building Excellence Award, Docklands Business Club Green Business Award 2015, MIDAS Best staff dining, MIA accreditation, and a certificate of excellence from Trip Advisor.

Since the building opened in 2012, the exhibition on Urban Sustainability has welcomed more than 200,000 visitors and 75,000 students, and has consistently achieved customer feedback in excess of 90%.

The event spaces have hosted hundreds of events, notably the Mayor's Low Carbon Prize, The Guardian Green Business Awards, Tesla Car Launch and the Green Apple Awards 2014. The Crystal has also published three books in conjunction with global experts, designed to engage readers in sustainable living at local, national and international levels.

WHAT DID THE PROJECT INVOLVE DOING?

Concept

The Crystal was driven by a vision of a multi-faceted urban world. Its crystalline shape represents the many facets of sustainability and the complexity of urban life. The Crystal has been specifically designed to be the world's most sustainable building and is one of London's newest landmarks.

Built in 2012, it has been created to start the conversation about sustainability through events, conferences and interactive exhibits.

WHAT IS INNOVATIVE ABOUT THE PROJECT?

The building is the epitome of innovation and demonstrates new solutions to existing problems through intelligence. The Crystal is a smart building and it showcases clear and tangible environmental benefits.
To quantify this, it is 42% improved consumption against a building with similar use. It achieves such a statistic through a variety of methods. For example, take natural ventilation. There are 150 ventilation panels which allow the building to naturally ventilate under the following conditions: if the temperature is between 19^{0}C and 24^{0}C inside and 20^{0}C and 25^{0}C outside and the wind and rain are at less than 5m/second and less than 35^{0} angle on a façade, then the building will naturally ventilate (open the ventilation panels).

As no fossil fuels are used on site to provide heating or cooling, The Crystal once again demonstrates its commitment to sustainability by a system containing two identical ground source heat pumps connected to a network of pipes buried in the ground under the building and the surrounding area. The total length of this pipe work is 17km and it provides a heat extraction of 600 kW which covers the total heating and cooling demand of the building.

The building was created by Siemens as a showcase for what is possible in terms of sustainable engineering, technology and operation. Siemens invested £30 million in the original project and currently use the second floor as an office space.

The building itself creates a profit by a number of different channels, such as the onsite exhibition, café and event hire. Since its opening in 2012, the exhibition has attracted more than 200,000 visitors, with the majority being paying customers. The building is also home to 12 flexible event spaces which bring in a large amount of business and have been used to for a variety of events, specialising in those with a similar sustainable message.

Siemens' employees are among the most frequent customers, showing their continued support of the building. Demonstrating their commitment to The Crystal's success, many senior personnel have visited the venue

Seeing the light: The Crystal's airy interior

and voiced their support, including Roland Busch, CEO, Infrastructure & Cities Sector, Siemens AG and Pedro Miranda, Corporate Vice President, Head of Corporate Development Siemens One, Siemens AG.

WHO AND WHAT HAS BENEFITED?

Environmental benefits

The Crystal has generated a clear environmental benefit. This is evidenced by its highly innovative infrastructure. The exterior is made entirely of glass – no two pieces are alike in size or shape. The glass is triple-glazed with an argon filled cavity for insulation and a low heat loss ratio. The exterior comprises six different types of insulated glass panels, each suited to their particular space and taking into account the angle of the façade and talking advantage of the sun's path. Different pieces of glass have varying degrees of transparency – clear, translucent and opaque (the walls at the end of the street are 64% opaque).

These measures ensure that the building does not overheat like a giant greenhouse but can also absorb the sun's heat when needed. Only 70% of daylight heat is allowed through, with 30% solar gain.

The Crystal has been designed to avoid the use of fossil fuels on site. Instead The Crystal is an all-electric building which will immediately benefit as the grid decarbonises over time. The venue has an energy demand of 1,467 MWh, 82.5% of which is provided from off-site sources including renewable power from the London Array, an off-shore wind farm in the Thames estuary. 17.5% or 256 MWh of the electricity is produced by a 1,580m_2 photovoltaic array mounted on the rooftop of the building, which provides 294 kW peak power – enough to cover the annual demand of the ground store heat pump. On-site production is required to achieve BREEAM and BREEAM certification.

Social benefits

The Crystal has produced various social benefits too. This has been achieved through actively educating and inspiring all about urban sustainability and the future role of technology in our lives. Offering free entrance to the exhibition for all education groups, our experts also host games throughout the year to further educate the younger generation.

The Crystal's workshops are held in our education suite and focus on sustainable methods to solve the problems we face now and in the future. As an example, one work shop tasked pupils to create a 'sustainable' object from materials found in the recycling bin. These activities attempt to not only engage participants in the subject of sustainability, but also to provide teaching techniques that centre on the freedom of thought.

Economic benefits

The Crystal has quickly become an integral element to the local and wider London economy. It has hosted a range of impressive events and international conferences. The Crystal celebrated its first anniversary by hosting Sustainable Cities Week. This included the G8 Innovation Conference with British Prime Minister David Cameron and London Mayor Boris Johnson's Low Carbon Awards, as well as a TEDx Newham event and the International Federation for Housing and Planning's 100th anniversary lecture.

The Crystal is also home to the Siemens Center of Competence, an internal group of sustainability, urban development and cities experts. The groups is responsible for building up strategic partnerships between Siemens and City organisations, conducting urban research, promoting cross-division and cross-sector solutions for cities, bringing together internal and external urban experts and managing the worldwide network of Siemens city account managers.

In doing so the Crystal is a front runner in educating all ages, ethnicities, nationalities and economic sectors on how they can adopt a sustainable lifestyle.

THE Crystal
A London venue with a point
Walking through our doors is like stepping in to the future; a future of greener, more efficient living. Greener, more efficient business. Reward your guests with an inspirational event space and innovative meeting rooms. Launch your product in the auditorium or gather a few choice people to develop an idea in a room with a picturesque view.
The Crystal is a purpose built conference centre, showcasing technology in a new perspective. The perfect choice of venue to get your point across and a truly unique opportunity to look at business from a different angle.
Give your guests a new perspective.
Book of one our 14 flexible event spaces today, all customisable to suit your businesses needs, hosting up to 720 guests. Enjoy our award winning facilities and reduce your carbon footprint at the same time.
To find out more visit us at thecrystal.org or call 0207 055 6400
A Sustainable Cities Initiative by
SIEMENS

TOSHIBA TEC UK

ENERGY SAVING

PROJECT AIM

The e-STUDIO 306LP MFP combines with the e-STUDIO RD30 for removing the toner colour from the printed documents. It is the world's first eco-friendly MFP system that can erase images and text on the prints.

The Toshiba eco-friendly e-STUDIO306LP/RD30 allows cost-saving yet has a positive green effect in the office today. Paper usage is reduced to one-fifth and provides support copy and print speed of 30ppm.

PROJECT ACHIEVEMENTS

The e-STUDIO can use paper up to five times and reduce CO_2 by 55%. Users can print and copy as they would on a typical MFP. However, when used paper is fed into the eraser unit, it heats the toner and de-colourises the print and provides a blank sheet to use again.

Just as importantly, each piece of paper can be reused over and over again with a conservative average reuse of five times, reducing paper costs and waste by 80%. This dramatically cuts paper usage and saves businesses money, so not only is it good for the environment, it offers an excellent return on investment.

WHAT DID THE PROJECT INVOLVE DOING?

The Toshiba eco-friendly e-STUDIO306LP/RD30 is a fully-fledged multifunctional printing system and eraser unit, which sets new standards in office waste reduction and cost savings. What makes this product unique is the use of a special blue toner, which can be de-colourised.

When used paper is fed into the eraser unit it heats the toner, virtually erasing the print and providing a blank sheet to use again. The same piece of paper can be used at least five times, dramatically cutting paper usage and saving businesses considerable money.

Green machine: paper can be reused five times

Caring for the environment and its resources is part of Toshiba's tradition. To contribute to a more sustainable society and as a globally operating enterprise, Toshiba gives highest priority to conserving and protecting resources during every part of a product's life-cycle.

Toshiba takes great measures to conserve energy and resources and reduce the burden on the environment. At the same time Toshiba understands the needs of its customers, ensuring its products and solutions not only have an ecological benefit, but also impact the economy and efficiency of businesses.

Toshiba's eco-friendly e-STUDIO306LP/RD30 printing system can erase images and text from paper. By using a unique heating technology this system removes the toner colour from printed documents.

The key features of this eco-friendly printer are its ability to sort paper into reusable and un-reusable sheets at the same time as removing the toner colour, and to digitise the documents prior to erasing.

Normally, "temporary" documents that have been used, for example, at a meeting, are either filed away or scrapped, but with this system, there is no filing or destroying of paper documents. The contents of the documents can be scanned and converted into a digital copy and saved for future use.

Based on the Toshiba strategy of Differentiation in ECO-printing, Toshiba has always provided multifunction products (MFPs) with enhanced environmental technology by reducing the MFPs size and weight and smart energy efficiency improvements.

A key cost-saving issue with a positive green effect in every office today is "saving paper". The e-STUDIO306LP/RD30 reusable paper solution opens a window to a new work style in the office environment by encouraging users to reuse paper through the new system.

In the long-term enterprises will be encouraged to adopt business strategies that contribute to sustainable development.

WHO AND WHAT CAN BENEFIT?

The ideal environment for the e-STUDIO306LP/RD30 would be a setting where sustainability is high on the agenda. Education and healthcare would be typical sectors that could easily utilise lower paper usage by up to 80 per cent, saving huge amounts of costs, as well as sustainable factors.

The e-STUDIO306LP/RD30 represents a massive leap forward in this respect, as standard copier paper can be used. Printing and copying functions operate as on a standard MFP, but when used paper is fed into the eraser unit it heats the ink, which wipes the print and provides a blank sheet to use again.

Just as importantly, if the same piece of paper is used at least five times, the total CO_2 emissions can be reduced by 60 per cent compared to printing once per sheet – dramatically cutting paper usage and saving businesses money.

ARE YOU PLANNING ANY FURTHER DEVELOPMENTS?

Since its launch in 2014 the e-STUDIO306LP/RD30 has been used successfully in several education establishments, as well as commercial and multi-national organisations. Toshiba have a big responsibility to help customers achieve a smaller carbon footprint. With the passion of leading innovation, Toshiba are constantly striving to enrich our reputation by introducing Toshiba e-STUDIO306LP/RD30 to more customers.

With this in mind, continued development is currently taking place on the e-STUDIO306LP/RD30. This development is based on customer feedback received to date.

TREETOPS, SINGAPORE

TREETOPS EXECUTIVE RESIDENCES DISCOVER GREEN PROJECT

PROJECT AIM

Treetops Executive Residences reduce our environmental impact by reducing our energy consumption and carbon footprint. Our corporate philosophy seeks to educate and promote the importance of going green to all our staff and guests by allowing them to get a feel of how easy it is to go green during their stay with us.

A new leaf: Treetops provides a greener living environment

The adoption of a green philosophy that inculcates healthy living into the daily lives of the guests at Treetops allows them to experience first hand and see the benefits of going green and saving the environment.

Treetops is also actively involved in green programmes by conducting sharing sessions and regularly host green tours for interested industry players and the media to showcase our current green initiatives, while urging more establishments to join in the green movement for a more sustainable future.

PROJECT ACHIEVEMENTS

Treetops has achieve savings of up to 50% on its electricity consumption since 2007 with the installation of green retrofits and the state-of-the-art Tri Gen System that traps waste energy to heat up domestic water and run the chillers for air-conditioning within the estate.

Water conservation is also on the agenda at Treetops with the installation of water-saving gussets, dual flush toilet bowls, allowing us to be certified as an Water Efficient Building by the Public Utilities Board of Singapore. Other Go Green initiatives serve to provide a greener living environment for all guests and staff while playing a part for our environment and the community at large. With this, Treetops has also won numerous Green Awards in recognition of its green efforts.

WHAT DID THE PROJECT INVOLVE DOING?

Launched in September 2000, Treetops Executive Residences is a 220-unit eco-luxury serviced apartment managed by DTZ Debenham Tie Leung Hospitality Management Services.

A luxurious sanctuary nestled amidst lush tropical foliage at Orange Grove Road, which is just minutes away from Orchard Road, Singapore's retail and entertainment hub, Treetops is the ideal retreat for the city dweller who wishes to escape the hustle and bustle of the city without giving up its conveniences.

For a holistic living experience, Treetops Executive Residences is the first serviced residences in Singapore to create an ingenious style of enhanced living environment that aims at including green features in the premises and inculcate healthy living practices.

WHY DID YOU DO IT?

With the increased attention on adopting a green lifestyle in today's society, Treetops understands that the focus should not be just the basic of providing comfortable accommodation and good services for our guests, but instead, to improve the good and comfortable living quality of our guests and staff while contributing in the conservation of the environment as a whole.

At Treetops, we understand that today's complex and high consumption lifestyle has always been a threat to the environment; these led to the kick-start of Discover Green project by becoming one of the first serviced residences in Singapore to create an ingenious style of enhanced living environment that harness our core competencies of supporting the community and environmental objectives and ensuring the general well being of our guests.

Haven of peace: the Treetops' garden

We believe that every eco-friendly effort, regardless big or small, has its impact on protecting the environment and creating a better living. Our clear direction of green consciousness and working efforts towards energy conservation and reduction of carbon footprint has strengthened our branding as a serviced apartment that allows guests to play a part in the conservation of the environment.

Treetops aims to incorporate the concept of eco-living into modern urban living, where our guests can be close to nature as much as possible with more than 60% of the estate landscaped with greenery.

There are also numerous green retrofits and green initiates implemented at Treetops such as the Tri-Gen System to reduce overall energy consumption by 50%, utilising renewable source of energy, and opting for only non-toxic cleaning materials for the establishment, all of which allow Treetops to be uniquely established as a "greener serviced apartment".

WHAT DID IT COST?

The funding for the project is heavily subsidised by the government and a shared savings incentive is in place to fund for the project that took five years to complete. On completion, it allows the operation of Treetops to be much more sustainable and reduce our impact on the environment to a large extent. With this, Treetops has managed to garner various green awards in recognition of the efforts in adapting green technologies within the property and across its business functions.

WHAT ARE THE PROJECT'S KEY DRIVERS?

1. Corporate Social Responsibility

Understanding the growing concerns on environmental issues faced in today's society, Treetops seek to impose our intrinsic belief of adhering to being a Corporate Socially Responsible organisation for a worthy global cause. Treetops also sets the bench mark as one of the leaders in adapting green initiatives in the hospitality industry in Singapore.

Environmental Conservation

With the desire to be an environmental friendly organisation as well as the engagement of energy conservation specialist, Treetops has achieve savings of up to 50% on its electricity consumption since year 2007 with the installation of green retrofits and the Tri Gen System that traps waste energy to heat up domestic water as well as run the chillers for the air-conditioning within the estate. Water conservation is also done at Treetops with the installation of water saving gadgets and dual flush toilet bowls, allowing us to be certified as an Water Efficient Building by the Public Utilities Board of Singapore.

Recycling program
Recycling bins for the recycling of papers, cans, plastic, are located in convenient areas within premises to allow guests to participate in doing their part for recycling.

2. To create an eco-luxury living experience and to promote green initiatives
For a holistic living experience, Treetops Executive Residences seeks to educate and promote the importance of going green to all our guests by allowing them to get a feel of how easy it is to go green during their stay with us. The adoption of a Green Philosophy that inculcates healthy living into the daily lives of the guests at Treetops allows them to experience first hand and see the benefits of going green and saving the environment.

As the overall well-being of their mind, body and soul are taken care of at Treetops, they will also be more open to bring their habits of going green back to their permanent home after their departure, thus spreading the green initiatives to people from all over the world.

All of these additional efforts allow Treetops to be uniquely branded as a " greener serviced apartment", attracting corporate clients and individual travellers who prefer to support properties with green initiatives.

Apart from making Treetops and our green efforts known to our guests, Treetops is also actively involved in green programmes by conducting sharing sessions and regularly hosts green tours for interested industry players and the media within our estate to showcase our current green initiatives and retrofits that we have, while urging more establishments to join in the green movement for a more sustainable future.

With Treetops as a successful example, we hope to promote the benefits of engaging in green technology and influence more organisations in going green.

WHAT AND WHAT HAS BENEFITED?

The Discover Green Project comprises the Green Retrofits at Treetops, Treetops Policies for Sustainability as well as Driving Green Initiatives for our Treetops Guests

1. Green Retrofits at Treetops
With a commitment to energy conservation and an ingenious approach to energy use, Treetops Executive Residences was able to save an astounding 50% of their energy over the baseline that was established in 2007. The retrofit was carried out in two phases with phase one completed in 2010 and receiving the BCA Green Mark Gold Award. Phase two was completed in 2012/13, garnering us the BCA Green Mark Platinum Award 2012 , a Green Technology Award 2012 and many other green accolades.

Works in phase one:
The air-conditioning equipment was optimized to operate efficiently with less hydraulic losses and improved heat transfer efficiency with the use of variable speed drives. The existing system was made more efficient, allowing Treetops to deliver the same comfort and same convenience for less energy.

Works in phase two:
A tri-generation plant was installed to supply power to the building and the waste heat generated is recovered to produce chilled water for the air-conditioning resulting in a more efficient air-conditioning system. Waste heat recovered is also used to produce domestic hot water in a centralised water heating system to replace more than 400 independent water heaters.

A vertical wind turbine, a first of its kind in a local service apartment, traps exhaust air from the chillers to generate electricity , which is in turn used to power lightings at the basement car park. In addition to the Green retrofits, Treetops also embarked on additional Green Touches to create a holistic all-comprising green enclave for our staff and guests

Treetops policies for sustainability

In a pledge of commitment to promote a green environment at its premises, Treetops Executive Residences listed in its Green Procurement and Sustainable Practices to make effort in adopting greener practices in its operation and minimize the harmful impacts on the environment.

a) Environment Policy

The environmental policy commits Treetops to carry out operations that are environmental friendly and also governs our criteria for making purchases and requisitions that have minimal environmental impact and engaging third party organisations that have demonstrated and implemented environmental preservation activities.

The basic components of this policy include:

- To produce less waste materials by reviewing how existing supplies, materials, equipments are manufactured, purchased, package, delivered, used, and disposed regularly
- To enhance energy efficiency by reviewing how current processes and systems can be improved such as installation of gadgets and overhauling the entire processes/systems
- To reduce wastage, improve recycling efforts through contractual relationships and purchasing practices with vendors, contractors, businesses, and other public and governmental agencies.
- In addition, the management encourages the purchase of products that have recycled content or more environmentally friendly, unless such products do not perform satisfactorily and/or are unreasonably expensive.

Go Green Day

Treetops has a Go Green Day since 2011, which aims to drive the message of caring for the environment through inculcating healthy living practices in daily lives so as to promote a greater sense of social responsibility.

This ongoing campaign, which is held on every alternate Friday of each month, serves as a constant reminder for the guests and staff that it is easy to play a part for the environment if we take steps in incorporating greener choices into our daily lives.

On Go Green days, we highlight and emphasises the three Don'ts:

- Don't drive – take public transport instead
- Don't take the lift – use stairs wherever possible
- Don't take fried food – opt for a healthy menu of fruits and vegetables instead.

In addition we run Discover Green, which is a series of emails sent regularly to keep guests updated about the green features added at Treetops while offering some health and "green" tips that can be easily implemented into their daily lives.

Many other green initiatives would be discussed in the detailed section of our Green Initiative. An digital copy is sent to reduce paper usage.

Green linen card

Our current housekeeping practice provides regular bed linen change on every alternate day. With the adoption of this new environmentally friendly initiative of water conservation, we seek our guests' continuous participation in the Green Linen Card where bed linen change is done twice a week. Participation levels had been high, where an average of 90% of the guests checking in have opted in.

Treetops recycling programme

Guests at Treetops are encouraged to recycle waste by disposing of recyclable materials including papers, cans, plastic, in the recycle bins. In addition to promoting recycling, the fundamentals of reduce and reuse such as reducing print and recycling paper in daily work lives are also undertaken. We see a year-on-year increase in the rate of participation for recycling and aim to do better.

All the cleaning materials used in the apartments are certified as environmental-friendly and non-toxic. All toiletries placed in guest rooms are 100% free of phthalates, sodium lauryl/ laureth sulphate, propylene glycol and parabens.

Lightings within the premises were replaced with energy efficient T5 and LED bulbs that emit less heat at the same time. In total 1,212 inefficient bulbs were replaced, while maintaining the same amount of illumination.

TRUSTFORD

TRUSTFORD RACES AHEAD ON GOAL TO BECOME A GOOD CORPORATE CITIZEN

PROJECT AIM

TrustFord is the world's largest, dedicated Ford dealer group, employing more than 3,000 staff. Operating across 65 sites around the UK poses a number of challenges for the measurement of energy consumption and the implementation of change.

As a retail brand providing the very best customer experience, TrustFord takes pride in its technologically advanced showrooms and also ensures face-to-face engagement with our nationwide customers. Therefore, energy consumption is a major cost for us, with energy bills well into seven figures. Our biggest environmental impacts come from the consumption of gas and electricity, transport fuel usage and the creation of waste.

TrustFord's strategy determines that it creates an environmentally conscious culture to reduce detrimental impact on the environment. Being a Good Corporate Citizen is one of only five overarching strategic goals set by board directors. Our work to become a more sustainable organisation forms a large part of this.

Our key environmental goals are:

- To reduce operational carbon emissions to a sustainable level
- To reduce water usage
- To reduce waste by reducing consumption, reusing resources and recycling waste
- To increase the number of BREEAM accredited properties to three by the end of the year.

PROJECT ACHIEVEMENTS

In 2014, we began to deploy a number of strategies to help us reduce our environmental impact. Now a full year since these strategies have been introduced, TrustFord is making strong head way and realising the goals set.

This success is largely due to dedicated, enthusiastic and skilled professionals who are passionate about helping us reach our goal to become an even better corporate citizen. For the year 2014, carbon emissions from gas and electricity use across mainland sites fell by 18% when compared with the same figure for 2013. This is an estimated 1,522 tonnes of carbon savings. Due to the ongoing encouragement of teleconferencing, TrustFord also made 12% year on year saving (May 2014 to May 2015) on fuel consumption by colleagues.

Some reasons for this include:

- Roll-out of an energy management system, initially to promote efficient behaviour
- Environmental champions being appointed in all sites, affecting behaviour among colleagues
- Installation of energy efficient lighting
- New buildings of higher environmental standard being operated
- Destratification fans being employed in a number of sites alongside changes to air conditioning systems
- On site oil burners
- Natural ventilation systems instead of air conditioning in showrooms.

WHAT DID THE PROJECT INVOLVE DOING?

In 2014 we began to deploy a number of strategies to help us reduce our environmental impact. Now a full year since these strategies have been introduced, TrustFord is making strong head way and realising the goals set. Some of the highlights of the initiatives we have implemented and their success to date are below.

Introducing a comprehensive Energy Management System

In 2013, a review identified areas of the TrustFord business where energy wastage was taking place, not only due to inefficient technology, but also human behaviour.

Following this, we recognised the need for sustained investment and a structured approach to measure our carbon footprint.

TrustFord appointed external energy specialists Purchase Direct to introduce a comprehensive energy management system. This allows us to set targets and implement energy reduction initiatives by measuring daily energy usage via smart meters. These record gas and electricity consumption at 30 minute intervals, feeding the data into useful graphs for the dealerships to interpret. Monthly reports are also sent to senior management to ensure long term goals are being met.

If our targets are missed, we analyse the graphs to understand why there is an unnecessary usage of energy so we can correct this immediately.

Within one year through behavioural change, TrustFord aimed to save up to 590 tonnes of CO_2. The actual amount we saved was 1522 tonnes of CO_2, a reduction of 18% compared with 2013. Our new target is 20% against this benchmark.

Putting our people at the heart of sustainability

At TrustFord, our staff are at the heart of our business. Therefore, it made sense for them to be at the heart our sustainability success.

In November 2013, we introduced a new initiative to recruit Environmental Champions. We engaged with colleagues across all of our 65 sites to connect with those who had a personal passion for the environment. An environmental coordinator was recruited into the business specifically to manage our environmental performance, with the skills to help give structure to our sustainability programmes across all sites.

We worked closely with Purchase Direct to ensure our environmental champions were trained to monitor and manage performance, identify appropriate corrective action and to help shift the attitudes of colleagues. Our champions have the full support of senior executives and General Managers at each site.

Sustainable new developments

As we operate in large workshops, some aisles are empty for much of the day. Therefore some new sites will have motion censored LED lighting that turn on only when needed. Electric radiant heat and solar panels are always considered across new developments. We also look to implement waste oil heating wherever feasible. While this is a more expensive investment, it allows us to be self-sufficient on fuel and reduce our carbon imprint.
A new development we are especially excited about is our Dagenham transit centre. Designed to achieve a BREEAM accreditation, it will boast rainwater harvesting, photovoltaic solar panels, energy efficient lighting and waste oil fired heating for the workshop. It will also have permeable car parking, rather than a solid tarmac, to help the water drain away naturally. This will be our third BREEAM development.

Our new parts warehouse in Warrington has been built using the existing steelwork of the current building, extended to raise the roof level rather than demolishing and building anew. This has saved both in steelwork and foundation work, both carbon intensive.

Use of waste oil burners

Four of our biggest workshops recycle waste oil from vehicles onsite. Typically, waste oil is removed from dealership sites by a specialist contractor creating risk of contamination. We have therefore invested in a special burning system that burns the oil at high temperatures to remove impurities, giving us a clean and cost efficient way of heating our workshops.

At TrustFord, we recognise the importance of upfront investment to achieve our sustainability goals. Using an example, the cost of a waste oil heating system for one workshop is 37.5% higher than a conventional gas heating system (£16,000 gas vs £22,000 for waste oil). In this particular instance, we also needed to upgrade the waste oil storage tank at a cost of £6,000.

We project fuel cost saving from the waste oil system of £3,000 per annum. Payback on the increased cost of the waste oil system would therefore be four years.

Rolling out energy efficient lighting

As our business requires modern showrooms with air conditioning and lighting, this is a large part of our energy consumption. Over the past two years, we have been working closely with Ford to review the lighting schemes in showrooms to implement low consumption lighting wherever we can. We're also seeking to replace showroom air conditioning with natural ventilation systems.

The TrustFord 'Lightbulb Moments' project was also rolled out in 2013 with the aim of reducing our lighting energy consumption, which can contribute to around 45% of a site's typical electricity bill. The project is still ongoing, and involves the upgrade of compact fluorescent workshop lighting to induction lamps.

Installing photovoltaic solar panels

Solar PV is currently being rolled out at Tamworth, Warrington, Kingston and Woodford Green. These were installed in the first half of 2014, and within two months have already delivered 14.2 tonnes of carbon savings.

Once all in place, energy consumption is predicted to be reduced up to 27%, saving collectively 62 CO_2 tonnes per annum. With 14.2 tonnes already achieved, we're on target to beat the goal set for this year.

Conclusion

At TrustFord, we are committed to improving our environmental position, and to become a good corporate citizen. We have invested in various initiatives to consciously take care of the environment and promote energy saving across all 65 of our sites, with great successes to date. As an expanding organisation, we have a program of new development ahead, with new projects all designed to be resource efficient and sustainable.

Moving into the future, we want to not only maintain the level of saving, but drive our consumption further and further down over time.

UAC BERHAD MALAYSIA

UAC BERHAD ENVIRONMENTAL INITIATIVES

PROJECT AIM

UAC Berhad Malaysia is committed to reduce the carbon dioxide emission from product manufacturing in order to reduce the environmental impact and to preserve the natural resources by the initiatives of optimise the electricity, fuel oil, raw material and recycled material usage in factory.

The main targets for the project are as follows:

- To reduce the carbon dioxide intensity of production from Scope Three emissions of raw material purchased
- 40% reduction in CO_2 emission intensity of production from the scope 2 emissions of electricity
- 20% reduction in CO_2 emission intensity of production from Scope One emission of medium fuel oil, MFO usage
- o optimise the raw material usage from product mix formulation study which reduce the usage of material with high CO_2 emission factor and to increase the percentage of recycled material usage without compromise on product quality standard
- To achieve cost saving with the reduction of electricity, fuel oil and raw material usage
- To review and explore the opportunities to reuse the scrap and sludge material in order to reduce the waste to the landfill which contribute to the Scope Three emissions of landfill waste.

PROJECT ACHIEVEMENTS

CO_2 intensity reduction:

- Scope One emissions of medium fuel oil has been reduced from 0.108kg in year 2008 to 0.099 kg of CO_2e per metric tons of product in year 2013, an 8.3% reduction
- Scope Two emission of electricity usage reduced from 0.306kg in year 2008 to 0.199 kg of CO_2e per metric tons of product in year 2013, a 35% reduction
- Scope Three emission of raw material usage has been reduced from 0.399kg in year 2008 to 0.388 kg of CO_2e per metric tons of product in year 2013, a 2.8% reduction
- Recycled pulp usage is increased from none to above RM2 million in year 2014. It helps to reduce the waste to the environment.

WHAT DID THE PROJECT INVOLVE DOING AND WHY DID YOU DO IT?

The project is expected to be planned and implemented in two different stages. The preliminary stage on the establishment of framework for monitoring and improvement has been completed and the achievements are as below:

- Baseline measurement for the electricity, fuel oil and raw material usage in term of carbon dioxide emission intensity of production for easier monitoring of the improvement result has been established in 2009. In the same year, UAC had registered with Institute Bauen Und Umwelt, IBU Germany for the environmental product declaration, EPD for cellulose fibre cement flat sheet with the EPD number of EPD-UAC-2010111-E and this has been revised in year 2013 to EPD-UAC-20130008 IAC1-EN
- Energy and waste management committee has been setup in order to establish the energy reduction and waste management objectives and target and to review the energy savings initiatives
- An energy manager has been appointed to carry out the plant wide energy audit and to propose the energy saving opportunities
- Carbon footprint reduction – CFP product certification obtained from national certification body, Sirim QAS.
- Major purchased raw material and utilities CO_2 emission factor has been established as below:

See chart on next page.

With the baseline data and the raw material and utilities emission data that has been established, monitoring of the CO_2 emission reduction and the effectiveness of the improvement program can be carried out in a more systematic & measurable manner.

The second stage of the project is related to the implementation of the initiatives identified in order to achieve the target CO_2 emission reduction.

Some of the main activities implemented are as below:

- A thorough energy audit has been carried out by the Registered Electrical Energy Manager appointed and opportunities for improvement has been identified for implementation
- Data collection to setup the baseline for various energy accounting centers, EAC
- 3Power factor improvement on all the high consumption motors such as refiner, ball mill and water pump
- Continuously improve energy efficiency at equivalent production level (kWh per production unit) through process and facilities optimisation such as VSD parameter optimisation
- Adopt new and efficient equipment & facilities such as purchase of a new more efficient boiler
- Progressively motor resize and replace with the high efficiency motor EE1 in order to improve the motor efficiency
- Reuse of scraps and rejected product for the fabrication of wooden strip and wooden pallet has been implemented
- Continuous research and development for the mix formulation optimisation has been carried out and a lot of trial run has been conducted to finalise the mix with lower CO_2 emission intensity, but still maintain and meeting the product quality standard such as MS1296:2010, EN12467, AS/NZS2908, BS476 Part6&7 and etc.
- Purchased of a new efficient boiler to replace the old unit in order to improve the MFO usage.

With all the implementation that has been done, electricity and fuel oil intensity usage has been reduced, raw material and recycled material usage has been utilised and directly this will help to reduce the CO_2 emission effect to the environment.

	GWP 100, kg / tonnes;m3;kWh;liter;km
Category / Item	UAC EPD declaration study
1. Cement Production	1057.66
	-515.66
2. Pulp production	
3. Silica Production	43.21
4. Alumina	159.07
5. Silica Fume	277.76
6. Red Pigment	1623.43
7. JKR water, per m3	3.75
8.Electricity, per kWh	0.87
9. MFO, per liter	4.05
10. Heavy diesel truck,per km	0.91
11. Ship Ocean, per km per tons	4.29E-09
12. LPG, per liter	1.62
13. Diesel, per liter	2.69
14. Disposal after use - solid waste to the landfill, per tonnes of FC	20

WHO AND WHAT HAS BENEFITED ?

Result comparison from year 2008 to year 2013 in terms of cost saving.

- Mix formulation review study that optimize the raw material usage: estimated saving RM5.4 million per year
- Electricity saving from the energy audit output: estimated saving 126 kWh per tons of product which represents RM 6.26 million per year
- MFO saving from the purchase of new boiler to replace the old inefficient boiler: estimated saving 1.99 litres of MFO per tons of product which represents RM 372k per year.

Result comparisons from 2008 to 2011 in terms of environmental impact reduction:

- Global warming potential (GWP 100 years) has been reduced from 776kg per metric tons of fibre cement flat sheet in 2008 to 673kg in year 2011, according to the EPD declaration report from IBU Germany.
- Acidification potential reduced from 3.9 kg of SO_2e per metric tons of fibre cement flat sheet to 3.305 kg
- Eutrofication potential reduce from 0.4 kg of PO4e to 0.387 kg.

Reduction of waste to the landfill:
UAC Berhad is the one of the pioneer company in Malaysia who registered the product EPD under IBU Germany. This EPD will work as a baseline or benchmark for other similar manufacturer in Malaysia. UAC had obtained several green certifications which are recognized worldwide such as Singapore green label, ECO Labelling, Carbon footprint CFP from Sirim QAS. Our product has been published in green pages directory Malaysia.

WHAT DID IT COST AND WHERE THE MONEY COME FROM?

The cost involved in the project is mainly from R&D on the trial of different mix formulation and also all the product testing for the certification. For environmental product declaration, UAC had to engage a consultant to work with the verifier in Germany. He carried out the reporting of our product EPD and it cost around RM70k.

For the R&D trial, UAC had spent more than RM 1 million per year. Besides that, UAC had spent more millions of RM1s or the purchase of anew efficient boiler in order to improve efficiency. The money for the investment is actually coming from all the cost saving from mix optimisation, electricity and fuel oil saving.

HAVE YOU ANY FURTHER PLANS?

- In the process of getting a new revised version of Gabi software to generate the carbon footprint data for UAC in monitoring the result more accurately. The cost is more than RM100k
- Purchase of a new ball mill and finishing cutting machine with higher efficiency to reduce electricity and labour dependants. These project shad been approve by board of director UAC and we are awaiting delivery. The cost is more than RM 20 million
- To convert the existing boiler from using medium fuel oil, MFO as heat source to using compressed natural gas (CNG). The cost is around RM200k but estimated saving would be more than RM500k per year
- Transaction of conservative motor to high efficiency motor
- LED light to replace the current metal halide in factory with the similar lux level
- To re-insulate all the steam pipe for steam loss reduction
- To study the feasibility of using cogeneration unit
- To study whether to using the OPC CEM II cement instead of CEM I with significantly lower CO_2 emissions. (DOE design)
- To evaluate the new type of raw materials such as lightweight additive to increase the volume of the output and this will indirectly reduce the total emissions of CO_2
- UAC is 50% through the completion of ISO 14001 which set out the measurable objectives and policy for environmental management system. It will help us to monitor the environmental performance in a more systematic and effective way.

WHAT ARE THE LONGER TERM BENEFITS?

UAC is among the first pioneer company who register environmental product declaration, EPD with IBU Germany. It set a benchmark for other industries to follow.

Besides that, by joining the pilot program on carbon footprint organised by Sirim QAS Malaysia who partner with carbon trust, UAC had obtained the CFP certification for one of our fibre cement product. It will provide a baseline for others manufacturer to benchmark.

Finally, with all the initiatives that carried out by UAC on improving the CO_2 emission intensity, we are in fact reducing the CO_2 emission intensity of production from 776kg CO_2 emissions/metric tons of product to 673kg in 2011 – a 2.9% reduction – and maintained at 683kg in year 2013.

ULSTER SUPPORTED EMPLOYMENT

TO ACHIEVE BS8555 STANDARD AT LEVEL 3

PROJECT AIM

The aim of the project was to find and implement a new accredited Environmental Management System (EMS), which would better suit the company needs. Like many organisations we introduce an EMS to satisfy a common condition in Government tenders.

We initially chose the Green Dragon standard as a vehicle for this but after a period of time and due to increased legislative and commercial requirements and an increased interest in environmental matters from our workforce, we found that Green Dragon no longer met our needs.

Research led us to decide that the BS8555 standard at level 3 would be more suitable for us. BS8555 is a much more formal system than Green Dragon. It is closer aligned to ISO 14001 but allows a phased implementation with an accreditation at each level.

One of the factors we had to take into account was our workforce. Ulster Supported Employment Ltd (USEL) employs a workforce made up mainly from people with a disability or long term health condition, and it is vital that we have employee engagement in implementing any new system. This provides training challenges as training materials need to be produced so as not to exclude for example people who are deaf or blind.

PROJECT ACHIEVEMENTS

The achievements to date have been many and varied. We have reduced our carbon footprint and improved energy efficiency by switching our water and space heating systems from oil to natural gas, producing a recognised improvement of 27% efficiency per KW/h.

By segregating our waste steams we have been able to reduce or waste to landfill by 60%. The products we segregate for recycling now include plastics, laser and ink cartridges, paper and drinks cans. As a by-product of this we have been able to help two local charities.

Employee engagement has also improved as a result of the initiatives, with employees starting to look for opportunities to recycle rather than throw away. To assist in this we have recently introduced an employee-led newsletter covering health, wellbeing and environmental issues.

Our intention for this year is to concentrate on company mileage both for business and commuter miles. And without the knowledge, experience and credibility gained in the implementation of BS8555, we would have been unable to go into partnership with our local council on the mattress recycling project

WHAT DID THE PROJECT INVOLVE DOING?

Ulster Supported Employment is a social economy company and an Arms Length Body (ALB) whose aim is to promote and create paid working opportunities for people with health related conditions.

To facilitate this objective we operate a factory in Belfast, manufacturing mattresses, divan bases and soft carrying products (bags). We employ 70 people in manufacturing, of whom 80% are disabled. A further 30 people are employed to provide training and support for disabled people in open employment, USEL supports approximately 1,200 people this way.

Due to the increase in environmental legislation that was impacting on USEL's business and the importance given to environmental matters in government and commercial contracts, the senior management that USEL should have a more formal approach to dealing with environmental issues.

USEL already had an Environmental Management System (EMS) accredited to the Green Dragon standard but it was felt this was no longer appropriate for the needs of the company. A senior manager was tasked with researching the different options available to do this and

come back with a set of proposals as to the best way forward for the company. From this, a decision was made that USEL should try to achieve BS8555 at Level 3 within four years. The BS8555 accreditation was chosen for a number of reasons:

- The standard allows an incremental approach with the differing levels needing greater complexity and structures, but we are still able to achieve an accreditation at each level
- Funding and advice was available to help with getting the initial accreditation
- While the Green Dragon standard was not robust enough for our needs other accreditation systems such as ISO14001 were seen as overly complicated for USEL's need at the present.

A gap analysis was commissioned in conjunction with an environmental consultancy to plan the phased introduction of the EMS. Three of the main findings were:

- The project would require significant commitment from management in the company
- Additional training would be required at all levels of the workforce
- Due to the special nature of our employees, additional resources would be required to achieve workforce buy in to the project.

We monitored all inputs with special attention to the energy needs of the organisation and once established, we reviewed the results, concentrating our efforts on reductions in areas which would have the most impact:

- We moved from oil heating to natural gas and employed a heating consultant to balance our heating system.
- We insulated the ceiling panels of the offices and issued all employees with a fleece.

WHAT DID IT COST?

The costs to do this were in the region of £4,500, which we used from our own resources with the expectation of a two year payback. This was achieved within 20 months.

Another project the company was able to justify from the analysis of records was to reduce the use of solvents in our silk screen process. This was done by introducing heat transfer printing which has virtually eliminated this harmful chemical from our premises.

This was a novel solution to the problem as previously transfer printing hadn't been possible on the materials we used, but the need to eliminate or reduce the use of solvents required us to reinvestigate the problem. This led us to find a company using a lower temperature transfer process, which not only worked with our materials but also gave us greater versatility in this area of our business.

In June 2014 we achieved the Level 3 standard one year ahead of the original target date. The following are the outcomes of the project:

- Energy reduced by 27% (from 920157 KW/H to 667744 KW/H)
- Waste generated reduced by 22% (from 201,566 litres to 157,300 litres)
- Water reduced by 19% (from 1591.5 M3 to1296.3 M3)
- Transport miles reduced by 4% (from 195,823 miles to 187,634 miles)
- Waste to landfill reduced by 40%
- All plastic milk cartons are recycled through a local charity
- Waste bins have been replaced by recycling stations
- All laser cartridges are recycled.

WHAT ARE THE LONG TERM BENEFITS?

This impressive reduction across a range of areas has reduced our cost base and assisted in the long term sustainability of the organisation. Other less quantifiable but equally important outcomes for us, included increased involvement with the community and worker participation.

We have developed links with two charities – the Play Resource Centre and Greenlight Gateway. We donate any raw material offcuts to the Play Resource Centre who then distribute these materials to schools and small community groups as art materials. All our drinks cans are donated to Greenlight Gateway, who crush them and sell them on for funds.

The involvement of our employees from the outset was a major contributing factor to our success and often it was them that pushed us further in our recycling efforts.

ARE YOU PLANNING ANY FURTHER DEVELOPMENTS?

Environmental issues are now reported through the company from employee forums to board level with some employees becoming environmental champions in their own communities. We undertake the Arena Network Northern Ireland Bench mark survey as a measurement of our position within the local environment and are presently at Gold Level and intending to move to Platinum.

Going forward, the knowledge and experience we have gained in achieving BS8555 Level 3 has pushed us into reviewing our production capabilities. As we manufacture mattresses and divan bases, it has persuaded us to develop a new production department working with our local council to recycle mattresses and divan bases.

This new section will remove approximately 3,000 mattresses per year from council waste streams with an on target recycle rate of 90% for a produce that currently goes straight to landfill.

VAUXHALL MOTORS LTD

BIODIVERSITY AND WILDLIFE IMPROVEMENT AND PROMOTION PROGRAMME

PROJECT AIM

The aim is to continuously improve and promote biodiversity and wildlife onsite, whilst educating our employees in the importance and benefits of maintaining and pushing a sustainable, diverse site and system.

Through this, we hope to keep our Wildlife Habitat Council certification and act as a major front against the idea of the manufacturing industry being an environmentally uninterested sector.

We have many important species onsite, and a historic background in the land we are based on, and through acting responsibly and sustainably we aim to respect and conserve this for generations to come. We hope to build on our ties with the Wirral Wildlife Trust and create further bonds with other national and international organisations, but our overall aim is to truly create a site ethos of environmental consciousness and pride in our naturally diverse place of work.

PROJECT ACHIEVEMENTS

The achievements of this programme and related projects are far reaching and varied. We have continuously gained Wildlife Habitat Council certification since 2010 through our efforts working for wildlife and biodiversity conservation and improvement.

We obtained funding to plant our own woodland, which is now well established and home to a vast array of species. A natural pond sits in the middle of it, whilst the trees provide a habitat for a multitude of songbirds, as well as

Still waters: a rain-fed lagoon is home to an abundance of birds and amphibians

Lesser Spotted Woodpeckers. We procured a peregrine falcon nesting box, which we now see chicks successfully fledge from every year.

Outside our head office we conserve a healthy colony of bee orchids, and within our woodland we have the largest helleborine colony on the Wirral. Our many wildflower, no-mow areas act as a huge boost to the local pollinator populations and our landscape garden onsite provides a habitat for solitary bees. Our site has also been communicated as a leader in environmental management and sustainability within GM and further afield.

Chicks away: one of the three peregrine falcons in the nesting box

WHAT DID THE PROJECT INVOLVE DOING?

We at Vauxhall Motors, Ellesmere Port take environmental management and sustainability seriously. We are always developing new ways to reduce, reuse and recycle any waste we produce, and continuously look for ways to lessen our impact on the environment.

Our facility was founded on a historic site; once the grounds of a stately home and then later a significant airfield in the world wars, it is now a car plant covering more than 350 acres. We have had the honour of building some of the most iconic vehicles in UK motoring history, and are now proud to be the home of the Astra.

The fantastic history and surroundings of our site give us a unique environmental profile which we feel compelled to conserve and promote, and work passionately to do so. We became the first manufacturing facility in the UK to achieve the ISO14001 accreditation, and now run the Biodiversity and Wildlife Improvement and Promotion Programme. After winning a Gold Green Apple award in 2004 for work on wildlife and biodiversity, we decided it was time we nominated ourselves once again.

The programme is an ongoing effort headed up by the site's environmental team with a basic aim to improve biodiversity on and offsite, and educate employees on the benefits of doing so– ultimately maintaining our Wildlife Habitat Council certification (we were the first manufacturing facility in Europe to be recognised with this accolade), whilst changing the perception of manufacturing as a less environmentally conscious sector.

These goals are achieved through a multitude of projects:

- Planting a woodland behind our paint unit with a tree representing every employee onsite. It is now well established with remarkable diversity; bluebells surround a natural pond, around which at the right times of the year you'll see and hear songbirds, lesser spotted woodpeckers and peacocks. These peacocks have previously been spotted nesting on the adjacent grassland we maintain to promote the abundance of wildlife that thrives in this habitat. In the woodland we also boast the largest helleborine colony on the Wirral as noted by the Wirral Wildlife Trust
- Designated no-mow areas around site in summer to allow wildflower growth, thus offering a stable habitat for pollinators
- A colony of bee orchids that we conserve and allow to flower freely year after year, right outside our head office.
- Successfully house nesting peregrine falcons every year, and this year we saw four chicks fledge
- A landscaped flower garden created onsite, with solitary bee nesting sites in the form of large purposed logs, and many flowers and grasses to promote pollinators. It also promotes mindfulness around why we work hard to have a limited impact on the environment
- Our rain-fed lagoon, once used as a storage lake, is home to many different birds, amphibians and mammals, and swarms of tadpoles in summer months. It's not

uncommon to see cormorants nearby, whilst shelducks swim on the water and buzzards fly overhead, all the while taking in the remarkable beauty of the area as host to a vast swathe of wildlife. Badgers, stoats, and otters have all also been seen on site and in nearby waterways.

WHAT DID IT COST?

The costs of these various efforts is hard to quantify as the materials used are usually recycled or repurposed. In times of austerity like these, obtaining a reasonable budget for matters that do not directly benefit production is difficult, but with a sound environmental mentality on site we are fortunate that we are able to cover any reasonable costs that sufficiently benefit and promote our environmental commitments.

WHAT ARE THE LONG TERM BENEFITS?

This programme is ongoing, so we are continually reaching and setting new goals. So far we have a site with outstanding biodiversity, and a mentality of conservation and sustainability.

The wildlife we have has gained us publicity and ensured we have an interesting side to our plant that always impresses visitors from many different nations, as well as auditors when looking at our environmental management. It shows we are going above and beyond our environmental obligations, and as such we have made a name for ourselves as a European leader in environmental management and sustainability.

he improved environment, we believe, aids in keeping Vauxhall Motors, Ellesmere Port a pleasant and interesting place to work, and in this way the programme has benefited the employees and business.

The future of the programme is bright. The environmental mentality of the site is ever-improving and the majority of the more natural habitats we manage have reached a stage where they require minimal input from us to thrive.

This means we have reached a stage where the promotion and education of the fantastic biodiversity we play host to can take the lead.

Woodland joy: another project

Our current project is the implementation of a Wildlife Information Board, to be placed outside the main site entrance, so employees and visitors alike can find out what they can expect to see on site, to create even more awareness of the importance of sustainability, and the role nature plays in the working environment.

Similarly, we are planning on rolling out smaller outdoor notice boards onsite so anyone can see what sort of wildlife may be in their immediate area, or what has been done in that area to improve the quality of the environment.

We are still looking to improve habitats, and at the moment we are constructing a large duck island to be floated on our lagoon, providing a safe nest site for the many aquatic birds that have made a home on and around the water.

We hope this application sufficiently conveys our passion and drive when improving environmental quality on and off our site. The work and effort we have put in to create a site where nature thrives alongside a large manufacturing operation is some of the best in Europe and there would be huge economic, sustainability and environmental benefits to the industry if this mantra were to take hold across the sector.

VECTAIR SYSTEMS

ENVIRONMENTALLY FRIENDLY AIR FRESHENER

PROJECT AIM

Vectair Systems is a manufacturer and supplier of hygiene products in more than100 countries worldwide. Vectair considers itself a true innovator in hygiene and its newest products are designed with environmental and social responsibility foremost in mind. Traditionally, aerosols or gel pots have been used to fragrance 'away from home' environments like washrooms, gyms and hotel reception areas. Aerosols are powerful and inexpensive and work well to provide a fresh fragrance where needed.

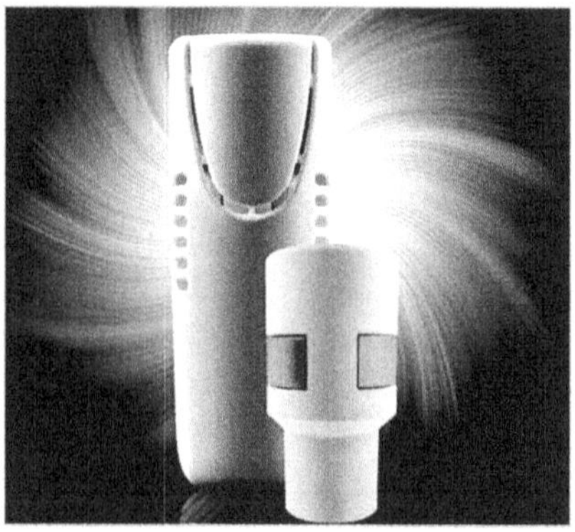
Breath of fresh air: V-Air® SOLID

Then came solutions without propellants – the chemical substance that 'propels' the fragrance into the atmosphere – utilising wicks or cartridge blocks that once heated, provide a pleasant fragrance. These are a great alternative but they use batteries or electrical power and Vectair wanted to go that one step further.

Vectair wanted to create a powerful air freshener that had as little impact on the environment as possible. It wanted to make a fragrance dispenser and refill that was VOC exempt and could be safely used anywhere for prolonged periods of time. It wanted to make a commercial air freshener that was 100% biodegradable and organic and a dispenser that was fully recyclable, meaning no batteries. This is how V-Air® SOLID was born.

PROJECT ACHIEVEMENTS

Vectair Systems manufactured and brought to market the V-Air® SOLID multi-phasing air freshener with sub-micron technology. It was officially launched in the UK at the Cleaning Show in March 2015.

V-Air® SOLID is classified as VOC exempt and waste-free. It is completely free of any solvents, propellants, HFCs and harmful volatile organic compounds (VOCs). V-Air® SOLID contains no batteries. It is designed so that it does not need any external power source such and it doesn't rely on gravity – all it needs to work is natural airflow. The refill is 100% organic and biodegradable and the dispenser is 100% recyclable. V-Air® SOLID utilizes pioneering sub-micron technology. The V-Air® SOLID particles enable superior fragrance delivery, yet are sized typically below 1.0 micron. This means they are much smaller and lighter than the particles in other fragrance systems, remaining airborne for many hours. This makes V-Air® SOLID highly effective whilst being safe for public areas and continuous exposure.

V-Air® SOLID is made up of a unique ceramic capsule and fragrance core. This, combined with a passive dispenser, delivers consistent fragrance intensity. The technology behind V-Air® SOLID is packed into one powerful component, with a retaining clip that attaches it to a dispenser.

WHAT DID THE PROJECT INVOLVE DOING?

Vectair Systems have been serving the cleaning and hygiene and facilities management industries for 25 years. Using their specialist knowledge, Vectair's product designer, managing director and other directors together came up with the unique concept of V-Air® SOLID.

They wanted to prove that cutting edge technology doesn't have to be technical to the end user – in fact, science can make things simple. This new innovation contains just three parts, yet has revolutionised the fragrance market.

WHY DID YOU WE DO IT?

Customers are increasingly looking to better their carbon footprint and so require more sustainable solutions. It is important however that these greener alternatives actually fulfil their primary purpose.

Vectair Systems looks to new technology to create greener alternatives while also focusing on the overall quality of each system and making sure it serves its purpose. Other systems on the market were experiencing problems such as leaks and they were susceptible to different climates. For example, they wouldn't work in humidity. They also weren't 100% environmentally friendly as they contain batteries. We wanted to make an effective air freshener for a commercial environment that was safe, totally recyclable and really worked.

WHAT DID IT COST AND WHERE DID THE MONEY COME FROM?

Vectair Systems itself invests the profit from its current successful solutions back into developing new innovations.

WHO AND WHAT HAS BENEFITED?

Establishments across the country can benefit as V-Air® SOLID is ideal for areas such as washrooms, offices, care homes, hotels, reception areas, school classrooms, and anywhere where fragrance is required. It can even safely fragrance confined spaces such as lifts and corridors, where this was not possible before.

WHAT ARE THE LONGER TERM BENEFITS?

The environment will of course benefit because of V-Air® SOLID's green credentials. Facilities managers will benefit because they can solve issues they once had with being able to safely and effectively provide fragrance to areas that they look after. Servicing companies will benefit from both sales, and cheaper servicing costs (no costly transportation or disposal methods, compared to aerosols). The end user will benefit from sub-micron technology (cleaner air, safer).

WAS THERE ANYTHING INNOVATIVE ABOUT THE PROJECT?

At Vectair Systems, we are exploring only the newest commercial aircare technologies to find innovative ways of controlling and removing odours. We have created an industry first – a multi-phasing, sub- micron air freshener delivering multiple fragrances from just one air freshener in a revolutionary way, and that contains no batteries, no solvents, no HFCs, no propellants and no aerosols. Because the V-Air® SOLID particles are typically below 1.0 micron in size, this means they are much smaller and lighter than the particles in other fragrance systems, remaining airborne for many hours. This makes V-Air® SOLID highly effective whilst being, very importantly, safe for public areas and continuous exposure.

CAN OTHER ORGANISATIONS BENEFIT FROM IMPLEMENTING YOUR METHODS?

Over time, in some locations you will find people can suffer from olfactory fatigue. This is the temporary, normal inability to distinguish a particular odour after a prolonged exposure to that airborne compound.

V-Air® SOLID ceramic capsules are cleverly infused with high quality perfumes, complemented by an internal fragrance core. These combine and diffuse powerful key fragrances at different evaporation rates. This means the start fragrance is different over time. This combats the issue of fragrance fatigue in locations where there is continuous exposure to fragrance materials.

Vectair Systems is a leader in its field and by creating these new and innovative environmentally friendly solutions, we hope to push the market forward and encourage others to look for new, greener ways of doing things.

WHAT DID YOU LEARN AND ARE YOU PLANNING ANY FURTHER DEVELOPMENT?

We are currently designing a brand new dispenser design that allows more air into the V-Air® SOLID cartridge to enhance airflow and therefore enhance fragrance intensity even further.

Since its UK launch at The Cleaning Show in March 2015, V-Air® SOLID has won The Cleaning Show Innovation Award – Washroom Category 2015, as well as an award in the USA – the Sanitary Maintenance Distributor Choice Award Winner 2015. This is voted for by distributors, not chosen by the magazine.

VICTORIA HOUSE PROPERTY PARTNERSHIP

VICTORIA HOUSE

PROJECT AIM

Our end goal is to achieve 80% recycling by the end of 2015, with new streams being trialled within the building, such as food recycling, clothing and batteries. We wanted to remove the use of black bags from within the waste streams and to establish cost savings on the overall waste management budget via improved recycling figures, a reduction in waste coming out of the building and finally via rebates for materials such as cardboard.

Site based cleaning and security personnel underwent training. The training reference NEBOSH Environmental Awareness at Work is a back to basics approach and provides the site staff with a basic knowledge on the need to identify with environmental issues in the workplace. The items that we are addressing at Victoria House are sustainability, pollution, noise, waste and emergency situations.

A target of 80% is regularly achieved and this data is collated by the waste company and distributed to all occupiers. A new glass waste stream was created due to the two retail and two events spaces that both have alcohol licences and predominantly sell bottled drinks.

WHAT DID THE PROJECT INVOLVE DOING?

The aim of our project was to review and begin establishing a market-leading waste and recycling process at Victoria House. This would take place in a number of ways, but primarily via re-fresher training, auditing and collaboration with key stakeholders such as occupiers and relevant service partners.

Our end goal was to achieve 70% recycled by the end of 2014, a target is regularly achieved. To start, the facilities manager underwent and passed the NEBOSH EMS Environmental Awareness at Work" course. We also set up a Green Forum with each of the occupiers feeding into it. The forum meets on a quarterly basis and regularly carries out table top exercises and we have regular guest speakers such as one from Bywaters, a waste management company. The companies based in the building have shown real willingness in buying into the project and have made changes in removing black bags from the building and replacing them with clear bags.

This allows our on-site cleaning team to conduct real-time audits of the waste from each tenant, allowing for any issues to be picked up immediately.

The building is now recycling between 80% and 85% and the management team and occupiers are really focussed on sustainability. Various items are under discussion, including:

- Purchasing of a waste sorting table for the recycling area
- Battery bin
- Food recycling
- Urban farm
- To link services with another building in the portfolio.

WHY HAVE YOU DONE IT?

We've made changes to our bin store area, which ensures that waste goes in the correct receptacle, and we have rolled out new signage for tenants to use, which leads to a better informed individual. The waste bin area is now clearly marked with colour coded lines with large easy to understand instructions/guidance, and with pictorial guidance for those for whom English is not their first language.

Victoria House is now committed to providing, as far as practicable, an economic and sustainable waste management approach resulting in a less costly use of economic and environmental resources.

Recycling of all waste where possible is an essential part of the sustainable waste management system at Victoria

House. The existing system is based on three main streams – mixed recyclables, general waste and glass, which accurately cover the majority of the needs of the building. The findings were in line with our beliefs in that there was plenty of scope to not only recycle more but also to reduce the amounts coming out of the building entirely, which is a key focus of any waste management strategy.

We took three main points from this audit, as follows;

- Around 30% of waste currently going into our general waste (non-recyclable) stream can actually be recycled
- The use of black bin bags was noted for general waste correctly, but also used for both recyclables and cardboard streams. The use of black bags automatically marks waste as non-recyclable
- 75% of our recycled stream is made up of paper and cardboard, with very little contamination, which is excellent.

WHAT ARE THE BENEFITS?

As a result, we now had factual evidence with which to engage our tenants and inform them of not only the highlights of the findings, but also how Victoria House's management team would look to address these issues in partnership with them. This included the following key points:

- Removal of all black bin liners for the building, which will allow cleaning teams to continue their good work of segregating recyclable and non-recyclable waste bags
- Renewed signage in bin locations
- Trial of food recycling with a view to further increasing recycled amounts coming out of Victoria House
- Relocation of large bins within loading bay area, thereby ensuring correct bins are used for waste and allowing for better management
- Recycling roadshows – based in our reception areas, these will engage with staff to become more aware of waste and recycling and the streams we have in place in the building.
- Creation of a Green Room – allow for smaller streams such as textiles and IT equipment to be collected.
- Organised tours of the recycling facility that we use, and a half-day visit to Nordic Recycling to better understand the recycling process
- Roll-out of refresher training of the on-site cleaning teams to update them on the new methods and to ensure they buy into the vision, recognising the key role that they play.

Resulting from these recommendations, positive changes have begun occurring in a short space of time, with the reduction of black bags being used, the relocating of bins and our loading bay manager overseeing the re-education of cleaning staff within the building.

Many of the measures which have been employed in the building have been at no cost.

VOLKER FITZPATRICK

WRABNESS EMBANKMENT STABILISATION

PROJECT AIM

We aimed to deliver a £1.25m drainage scheme to a section of railway embankment situated within an environmentally challenging and sensitive location.

The works at Wrabness have been required for more than years but had not been achieved due to the difficult environmental and ecological constraints. In previous years the embankment was close to a deep seated slip circle failure, i.e the effective collapse of the embankment and closure of the railway line.

We had five key objectives in the delivery of the project:

- To construct a drain system to alleviate the pressure on the embankment and to minimise the impact on the local community caused by disrupted passenger and freight rail services
- To minimise the short-term impact on the woodland habitat and the visiting public during the construction of the drainage scheme
- To leave a long term legacy for the ancient woodland environment for the benefit of generations to come
- To have a strong stakeholder engagement and involvement from the start incorporating their needs, concerns and aspirations into the overall scheme
- To put environmental consideration to the core of the design and implementation driving innovation and delivering an exemplary project.

PROJECT ACHIEVEMENTS

We have successfully delivered a drainage scheme within one of the most environmentally challenging and sensitive locations on the railway network. Our engineering solution will help protect the embankment for a further 100 years.

During construction, we introduced innovative measures to protect the habitat especially through protecting the ancient coppice stools and tidal estuary and to that the indigenous fauna occupying the woodland were protected

Wood work: ancient woodland has been enhanced

from harm. We worked with the woodland custodians, RSPB and devised a long term strategy to help develop and enhance this ancient woodland and have been in a fortunate position to be able to fund these improvements..

We have achieved a level of engagement with the stakeholders above and beyond what would be normally expected which has enabled the works to go ahead having been stalled for over ten years. The testimonials attached to this application highlight the exceptional outcomes we achieved for the stakeholders.

Uniquely, an Environmental Manager was embedded within the design and delivery team to drive environmental innovation from the beginning and into every aspect of design and execution. The final scheme will blend back into the environment as the woodland quickly regenerates.

WHAT DID THE PROJECT INVOLVE DOING?

This section of the Anglia railway network consists of a two- way track supported on a man-made embankment, built in the mid 1800s. The line has a vital role in reducing the impact on the local road network and a closure due to embankment failure would have cost the local economy millions during the emergency works to re-open it.

The adjacent ancient woodland has an undulating topography with an overall gradient towards the embankment Ground and surface water pressure has caused lubrication within the embankment causing the failures. The drainage scheme inserts 500mtrs of French drain style formation to intercept the water and direct it through pre-existing culverts draining to the estuary.
he history of the ancient woodland has created the unique and diverse habitat associated with ancient coppice woodland and why dormice are well established as are great crested newt and several rare moth species.

WHO AND WHAT HAS BENEFITED?

Environmental consideration was at the heart of both design and implementation of the scheme. There is a pre-existing problem where parts of the woodland are 'drying up' causing tree mortality and concerns were raised that the drainage improvement could make the drying process worse. Working with RSPB we have been able to put together a ten-year improvement plan which we are funding and which will be delivered by RSPB. The plan covers the following area's of improvement to reduce the impact of climate change to the woodland.

Works to significantly decrease the dominance of sweet chestnut as a coppice species to improve the general biodiversity and diversity of species within the woodland, This will be achieved through removing a percentage of the sweet chestnut coppice stools and planting a range of local tree species.

Ride Management mimics the historical management processes where they would have kept tracks or rides open of vegetation. Our process will include felling trees alongside the paths allowing in light and maintaining a relatively sheltered micro-climate. The vegetation will be cut annually or bi-annually, mimicking the grazing process.

A re-wetting program will see the installation of a dozen dams, with levels set to maximise winter rainwater storage. The result will be an increase in biomass and a diversification of invertebrates, plants and trees

WHAT WAS INNOVATIVE ABOUT THE PROJECT?

The inspection chamber design was modified at an increased cost with the benefit that the understory (undergrowth) could be allowed to envelop the metal lids. This will minimise the visual intrusion of the chambers for woodland users while allowing future detection by metal detector for maintenance purposes.

A modified 1.5km access haul route (1 km longer than the obvious direct route) ensured that the most sensitive parts of the habitat were avoided. There was potential for extensive damage to ancient coppice but a method was devised which involved installing a stool protection system which could withstand the weight of traffic movements preventing damage.

Careful consideration was given to preventing pollution or silt discharge to the estuary. A three-stage intercept system was temporarily constructed using a combination of silt filtration and "pollution event" containment. The site cabins were of an "eco-specification" with low energy consumption, including gas power generation as opposed to diesel. Water for toilet flushing and dust suppression was sourced from a harvested rainwater store on the farm adjacent to the site.

The site had stringent biosecurity measures in place ensuring that all personnel footwear and plant/equipment was disinfected before entering/leaving the site to prevent disease spread such as Ash die-back

Extensive recycled material was sourced locally to minimise haulage distance. At the conclusion of the works, the majority of the haul-road material was utilized by local landowners to improve access lanes in and around the woodland significantly reducing our haulage carbon footprint. No construction waste was sent to landfill from the project.

An Environmental Manager was embedded with in the design management team to ensure that the design and construction incorporated sustainability as a core value.

The site supervision team received training on site specific flora and fauna far above the standard levels of knowledge briefing to ensure that they were empowered to drive the environmental improvements forward throughout the construction.

Every individual accessing site whether they were visiting or working on site received a full dormouse licence and environmental control briefing.

It was extremely important to alleviate the public concerns. A public interface operative was permanently employed on site to assist the public in navigating around the works, providing advice and information about the works and the various ecological measures in place to protect the historical woodland.

Before any works were undertaken, the engagement with the stakeholders began at a concept stage and continued throughout the project. Regular hold points were established into the program for inspection by the stakeholders to maintain their confidence in our approach.

WHAT ARE THE LONGER TERM BENEFITS?

The enhancement works will provide a legacy for the public users which will ensure the ancient woodland flourishes for generations to come. A failure of the railway embankment would have a significant negative impact on the local economy. We were able through demonstrating very high environmental standards negotiate access with the stakeholders and undertake the works ensuring the railway continues to contribute and not hinder the development of the local economy.

We have established an extended period of monitoring in the habitat following the works. Although early days there is already evidence that the impact of our approach shows a regenerated habitat for dormice and the start of the longer term benefits for the wider woodland.

ARE YOU PLANNING ANY FUTURE DEVELOPMENTS?

Future works to improve the stability of the opposite embankment will need to be undertaken. We have already started species specific monitoring to fully understand the habitat on that side of the embankment and by taking forward the lessons of this project, that project will aim even higher for overall environmental benefit.

The level of stakeholder engagement was absolutely key to successful delivery and the project approach has been used as a case study so that other projects can learn lessons and repeat this success.

VOLKERHIGHWAYS

LB HACKNEY CIVIL ENGINEERING TERM CONTRACT

PROJECT AIM

A key aim of the contract is to deliver high quality sustainable services to the local community. It is also to support and integrate with the local community, tapping into the ethos of the local authority's objective of improving sustainable transport and carbon reduction.

This aligns with the objective of engaging with all parts of the community in a fair and indiscriminate way, giving opportunities to all and maximising the chances of those from different backgrounds and abilities. The project also aims to improve the service provided to vulnerable users within the community and continually strives to engage with these groups to understand issues and then working with them to identify solutions that improve their use of the highway service.

PROJECT ACHIEVEMENTS

Our dedication to maintaining good community and client relations have set a precedent for maintaining a high level of sustainable standards in future construction works taking place in the area. This is especially so in the areas of sustainable transport, community engagement and scheme delivery. Specifically we have:

- Introduced cycling inspections on a borough network – we believe we are one of the first companies to do so in the UK
- Completed an innovative "mystery shopper" exercise to understand issues facing disabled people using pedestrian management around our works
- Constructed one of the largest shared space junctions in the UK giving pedestrians priority use whilst offering users a unique approach to travel and transport user integration.

Also, the contract has set the precedent for eco-cabin best practice for VolkerHighways business. Eco-cabins use photovoltaic panels to generate power. Our team works closely with our cabin suppliers, providing feedback on our requirements for energy capacity, power storage and improved efficiency, so that we can help them to make improvements to their products. As a result of this, it is now a best practice on the LOHAC (London Highways Alliance Contract) for TfL.

VolkerHighways

WHAT DID THE CONTRACT INVOLVE?

VolkerHighways has been renewing and maintaining the highways and footpaths in Hackney since 2004, including footway and cycleway renewal and maintenance. We provide a range of reactive and minor improvement services, as well as major projects across the borough, completing around 5,300 task orders per year.

A key project was Leonard Circus, a dramatic change to an existing T-junction to create east London's first example of a 'shared space' street. By removing all signage and traffic lights, planting mature trees, and using high quality granite paving, we created a new junction that brings together pedestrians, cyclists and vehicles in a harmonious environment.

The scheme supported the council's transport vision and strategy of encouraging walking and cycling in the local area. It has made the junction more accessible for elderly and disabled people, and pushchairs. Plus, its reuse of existing materials helped minimise the environmental impact of the works (for example, the team reduced carbon by laying 315m2 of reclaimed, locally-imported York stone, reused from Old Street Station).

Hackney is a busy and congested area, so our team used a fleet of bicycles to travel between the various sites. This helped to reduce emissions and congestion, as well as support our client's sustainable vision for the local area.

Our team also fully engaged with the local community, including employing local people as apprentices, and working with local people with disabilities to ensure that our works did not cause obstructions to pavements.

WHY DID YOU DO IT?

The council's aim for the contract is to create and maintain roads and footpaths that everyone who lives in, works in, or travels through Hackney can benefit from. They also want more people to walk, cycle and spend time together safely in the community.

Streetscene: Leonard Circus creates harmony

Our works are an integral part of Hackney's on-going public realm 'streetscene' improvements across the borough – improving the visual quality of the area and making it a greener and safer place for everyone.

WHAT DID IT COST?

The contract value is around £10 million to £14 million per annum, over a four-year contract period, including £3 million of central government funding.

WHAT DID THE PROJECT ACHIEVE IN TERMS OF SUSTAINABLE DEVELOPMENT, ECONOMY, ENVIRONMENT AND/OR EQUITY?

The contract has employed two new apprentices each year since 2011, the majority of which come from Hackney. Around 40% of our supply chain's workforce live within ten miles of Hackney.

Using bicycles to travel between sites (approximately 1,500 miles) has resulted in a saving of 0.46 tCO_2e. Less than 1% of waste from the contract has been sent to landfill. Using 'ecocabins' with photovoltaic cells has saved 5,265 litres of diesel a year, equating to 13.6 tCO_2e.

WHO AND WHAT HAS BENEFITED?

The improvements to roads and pavements make it easier for people to get around, thus creating a more pleasant and attractive environment in which to live and travel through. By choosing to travel around the borough using bicycles instead of vehicles, our team helped to reduce congestion on local roads, and the associated emissions.

The contract has employed two new apprentices each year since 2011, the majority of whom come from Hackney. We support our apprentices in finding suitable work when their apprenticeship ends. For example, we are training one of our apprentices to become a supervisor on our upcoming contract to deliver the Cycle Superhighway through Hackney. By offering apprenticeships and employment to local people, and utilising the local supply chain, we have helped to improve the level of skills and training in the local area, and support Hackney in its drive to reduce the 'economic polarisation' that exists between the richest and poorest people living in the borough.

One of the biggest social benefits of the scheme is that vulnerable people in the area have benefited from the improved pedestrian areas. Creating streetscene areas such as Leonard Circus has helped to create harmony between pedestrians, cyclists and motorised vehicles in the area, and our work to improve highway junctions has helped to reduce congestion and its associated stress and frustration.

WHAT ARE THE LONGER TERM BENEFITS?

The main longer term benefit that this contract will leave is a legacy of skills and training in the local area, helping to close the skills gap that Hackney experiences, when compared to the wider London area.

WAS THERE ANYTHING INNOVATIVE ABOUT THE PROJECT?

Using bicycles to travel around the borough was an innovative way to reduce carbon, as well as promoting better health for our team.

Working with Disability Hackney – a local community group representing disabled people in the area – we arranged for 'mystery shoppers' to traverse our pedestrian traffic management and provide us with feedback on any accessibility issues for people with disabilities. This proved to be a very valuable exercise in improving our pedestrian management and increasing awareness and consideration for people with mobility issues.

After receiving feedback from local resident Mrs Wiles, who was concerned that construction warning signs placed on narrow footpaths might make it difficult for people to get past, we placed the signs on lampposts.

This provided unobstructed access to footpaths, while motorists and cyclists benefited from the increased visibility of the signs.

CAN OTHER ORGANISATIONS BENEFIT FROM IMPLEMENTING YOUR METHODS?

The success of our bicycling initiative has resulted in one of our joint venture contracts for TfL purchasing a fleet of bicycles for their team to get around their sites. Two bicycles have also been purchased for supervisors on our recently awarded contract to deliver the Hackney section of London's cycle superhighway.

The DfT has praised the potential of our innovative solution of placing roadwork signs on posts, and the next step is to roll out the scheme to other highways term maintenance contracts in London, where suitable.

WHAT DID YOU LEARN AND ARE YOU PLANNING ANY FURTHER DEVELOPMENT?

The team has experienced a lot of positivity and enjoyment from travelling around on bicycle, and have found that riding around Hackney together has actually helped them form a stronger team.

Pedal power: bike initiative has been a big success

Mrs Wiles was happy with our innovative approach to providing unobstructed footpaths during roadworks. More praise came from Quaysoor Miah, Principal Highways Engineer, Hackney Council, who said: "We have had an excellent service provided by VolkerHighways... We have received several compliments, both internally and externally, for schemes carried out in the ...borough."

As well as the obvious health benefits, bicycling through the local area, without the physical barrier created by a car or van, meant that they felt more immersed in the local area and the community. They could stop and chat to people, which would have been almost impossible had they been behind the wheel.

WYKE FARMS

WYKE FARMS 100% GREEN

PROJECT AIM

Our aim is to be one of the greenest brands in grocery today. Our '100% Green' plan is backed up by promises to significantly reduce our carbon footprint and to produce all our energy from our renewable solar and biogas sources. We are the only dairy company to make a list of green promises on our website that we measure ourselves against to monitor our progress and will publish on line.

These promises are:

- To produce all our own electric and gas from renewable sources
- As much as possible we will offset our third party supplier's energy usage
- Save energy through investment and changes in work practices
- Maximise use of organic nitrogen on our farm and supplier farms, replacing artificial fertiliser
- Minimise package waste
- Recover heat in our factory
- Filter and reuse waste water
- Knowledge share and encourage our suppliers to work in more sustainable ways
- Encourage wildlife within the Brue Valley.

These promises permeate throughout our organisation from marketing and product development through to production and communications with our customers. It even encompasses construction, as our water recovery plant was built with a 10-degree south-facing roof so that we can retrofit solar to run the plant on daytime power.

PROJECT ACHIEVEMENTS

- Made us 100% self-sufficient in our own gas and solar power – Fitter for the Future
- Saved costs in energy consumption (approx £2 million) and in packaging costs (reducing waste)
- Biogas plant – fuels our electricity-generating CHP plants and saved over five million kilos of CO_2 per year.
- Solar power – produces 80,000 kWh of energy per year and saves more than 42 tonnes of CO_2
- Water recovery plant – £1.3million investment allows 95% of our waste water to be reused, reduces factory usage by 70% and saves up to 850,000 litres per day
- Digestate: a by-product from our biogas plant is used as organic fertiliser, displacing artificial fertiliser use.
- Electric vehicles for local deliveries as they have zero CO_2 emissions
- Exporting upgraded biomethane gas back into grid – 6.6 million cubic metres per annum, saving 16 million kilos of CO_2 per year
- Reduced waste – farm and dairy waste is now utilised in an AD plant, resulting in zero road miles
- Packaging – new carton wrap machines reduced card weight usage by 25%. Changing working practices in production reduced packaging waste by 70%
- Product – invested £1 million in cutting equipment minimised waste, with block yields improved.

WHAT DID THE PROJECT INVOLVE DOING?

From an original 3,000-plus business, Wyke Farms is now one of fewer than ten farmhouse cheesemakers remaining in the West Country. Today we are the largest independent cheesemaker in the UK (in a cheddar category dominated by multinationals and PLCs), selling more than 13,000 tonnes of cheddar per year in the UK alone, and more to another 150 countries worldwide.

Wyke Farms brand is the third largest brand and the only independent company trading in the top ten brands. This success is down to many factors, one of which the family's commitment to sustainability and to operate our business in a way that has minimal impact on the environment and cares for the people involved in the business at all levels.

Our sustainable approach to business underpins the shoppers' confidence in our range of dairy products and helps to give shoppers the trust that they need. As part of our commitment to 100% green, we are one of only a very few food companies to make a list of green promises on our website that we measure ourselves against every year and will publish on line.

Through our 100% Green strategy and our Fit For The Future Strategy, we have set a target to be one of the most sustainable grocery suppliers in our sector. Through Fit For The Future, we are investing in state of the art technology and advertising and selling our brand across the UK and to more than150 countries across the world.

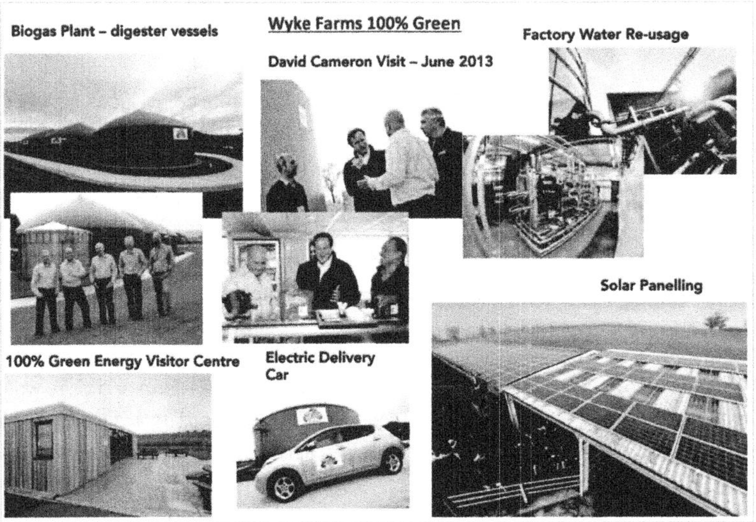

Big cheese: David Cameron got the inside story of the farm

This efficiency combined with out sector-leading quality enables us to create an added-value vehicle for our cheese so that our family can still be making cheese in this region for another 150 years, and the farmers in this region can continue to produce milk and paid a sustainable and affordable price.

WHO AND WHAT HAS BENEFITED?

We are also developing 100% Green Further: the third phase of our AD project, which is already under way and will generate more upgraded biogas, which will be used to provide fuel for HGV, including our milk tankers.

In terms of solar, we plan to increase this tenfold by utilising other roofs in the business and also recover heat off the rear of our solar panels and of batteries on charge. The second phase of our Water Recovery Plant has already started. To boost our resilience to our changing climate, we are collecting water in a 'drought pond' that is used for irrigation, and to improve the ecosystem. We have gained recognition for business sustainability performance shown by a series of accolades gained over the last year. We have been benchmarked against other suppliers in grocery and The Grocer Magazine awarded us a gold award for the most Sustainable Supplier in Grocery 2014.

Waitrose John Lewis partnership benchmarked us against all of their suppliers as part of their Waitrose Way Awards and we were the only supplier to win two awards in 2014, for water re-usage and carbon reduction.

With any Green energy project we think that 'natural fit' and a holistic approach is most important. It is not about shoehorning 100MW solar into prime agricultural land or about an AD plant on a business park with waste hauled in for miles and out for miles – the Fit for Green needs to be better than that.

In terms of challenges, we have faced all the usual things, including endless E.A. form filling and bureaucracy, planning issues and grid supplies for power and gas, which are not untypical of such a project.

One issue was processing our lactose waste as a feed for the AD that initially proved a problem as it can turn the process acid. We successfully developed a system using our waste pig slurry as a buffer, which now allows full usage of our lactose as well as waste from other dairies.

WHAT DID IT COST?

Securing finance was also a challenge, particularly for a private business such as ours. We were the first AD project and the first gas-to-grid project that Barclays had supported, raising around £8 million in total. We believe that this has also helped pave the way for other companies looking for finance for AD projects, as Barclays now acknowledges the value in green and has used as a case study in their high-level print (Financial Times) and television ad campaigns.

CAN OTHER COMPANIES BENEFIT FROM USING YOUR PRACTICES?

In terms of raising awareness of & encouraging sustainable practice amongst other businesses and communities, we have set up a visitor centre at our green energy plant so that we can help to share best practice with farmers, national retailers and educational schools and colleges. To date we have had more than 2,000 visitors.

We are also very keen on supporting others to reduce their waste & their carbon footprint. We are encouraging farms and fellow business to use solar and AD renewables. With local food and drink businesses, we have partnerships to take in their waste and process it to go back as digestate to their farms - help reduces waste and reduces artificial fertiliser usage.

Local farmers and contractors also use the digestate from our biogas plant to reduce their artificial fertiliser usage. In total saving 1.5 million kilos of CO_2 compared to using artificial fertiliser.

We also continue to raise consumer awareness of importance of sustainability through our PR, marketing communications and social media programme highlighting our 100% Green initiative and emphasising our short supply chain. The result has been widespread national coverage reaching over 50 million readers. We also focused on branding and developed a new logo: "100% Green" representing our commitment to manage our business sustainably which is included on our packaging and all related communications.

We are proud to be the first national cheddar brand to be self-sufficient in green energy and one of the greenest brands in the grocery sector today.

Made in the USA
Monee, IL
08 November 2026

67950723R00177